HOUSEHOLD CAPITAL AND THE AGRARIAN PROBLEM IN RUSSIA

Household Capital and the Agrarian Problem in Russia

DAVID J. O'BRIEN
University of Missouri

VALERI V. PATSIORKOVSKI
*Institute for the Socio-Economic Studies of Population
Russian Academy of Sciences*

LARRY D. DERSHEM
University of Missouri

LONDON AND NEW YORK

First published 2000 by Ashgate Publishing

Reissued 2018 by Routledge
2 Park Square, Milton Park, Abingdon, Oxon OX14 4RN
711 Third Avenue, New York, NY 10017, USA

Routledge is an imprint of the Taylor & Francis Group, an informa business

Copyright © David J.O'Brien, Valeri V. Patsiorkovski and Larry D. Dershem 2000

All rights reserved. No part of this book may be reprinted or reproduced or utilised in any form or by any electronic, mechanical, or other means, now known or hereafter invented, including photocopying and recording, or in any information storage or retrieval system, without permission in writing from the publishers.

Notice:
Product or corporate names may be trademarks or registered trademarks, and are used only for identification and explanation without intent to infringe.

Publisher's Note
The publisher has gone to great lengths to ensure the quality of this reprint but points out that some imperfections in the original copies may be apparent.

Disclaimer
The publisher has made every effort to trace copyright holders and welcomes correspondence from those they have been unable to contact.

ISBN 13: 978-1-138-72356-6 (hbk)
ISBN 13: 978-1-138-72353-5 (pbk)
ISBN 13: 978-1-315-19297-0 (ebk)

Contents

List of Figures		*vii*
List of Graphs		*ix*
List of Tables		*xi*
List of Contributors		*xix*
Preface		*xx*
1	Changes in Russian Agriculture: New Institutional Structures in the Countryside	1
2	Peasant Households and the Transition to a Market Economy and Democracy	15
3	Population and Household Structure in Rural Russia	43
4	Research Design	61
5	Household Structure and Labor	79
6	Community Attachment and Social Networks	93
7	Physical Capital	109
8	Household Agricultural Production and Sales	131
9	Employment, Household Income and Durable Goods	161
10	Mental Health and Symptoms of Depression	191
11	Subjective Quality of Life	213
12	Supporting Sustainable Households and Communities in an Era of Globalization	237

Appendixes	247
Bibliography	273
Index of Authors	281
Subject Index	283

List of Figures

Figure 1.1:	Institutional Structure of Rural Life in Soviet Russia	5
Figure 1.2:	Institutional Structure of Rural Life in Contemporary Russia	7
Figure 2.1:	Conceptual Model of Possible Direct and Indirect Effects of Household Labor, Social Networks and Community Attachment on Peasant Household Production and Sales	33
Figure 2.2:	Conceptual Model of Possible Causal Effects of Household Labor, Social Networks and Community Attachment on Household Income and Acquisition of Durable Goods	37
Figure 2.3:	Conceptual Model of Possible Effects of Household Capital on Subjective Quality of Life and Depression	38
Figure 4.1:	Village Location	64
Figure 8.1:	Structural Equation Model of the Combined Effects of Different Types of Household Capital on Peasant Household Agricultural Production	145
Figure 8.2:	Combined Effects of Different Types of Household Capital on Household Agricultural Sales	156
Figure 9.1:	Effects of Household Labor, Helping Networks and Community Attachment on Peasant Household Monetary Income from 1995-1997	179
Figure 10.1:	Modeling Household Capital and Symptoms of Depression in the Three Russian Villages	203

Figure 11.1: Effects of Household Capital on Five Quality of Life Domains and Life in General — 232

Figure 12.1: Future Institutional Structure of Rural Life in Russia — 241

List of Graphs

Graph 2.1:	Percentage Representation of Agricultural Producers, Land Used, and Mean Percentage of Commodities Produced by Large Enterprises, *Fermery* and Household Plots in Russia (1996)	21
Graph 2.2:	The Size of the Rural Population in Russia from 1989 to 1997 (in millions)	31
Graph 3.1:	Changes in the Size of the Rural Population, Number of Rural Households and Average Size of Rural Households in Russia (1960-1994)	50
Graph 3.2:	The Average Number of Households and Residents in Three *Oblasts* Where the Study Villages Are Located	53
Graph 3.3:	Five Types of Age Structures in Rural Russia	56
Graph 8.1:	Predicted Amount of Agricultural Production (kilograms) by Weighted Number of Adults	138
Graph 8.2:	Number of Nonredundant Ties and Predicted Household Production (kilograms)	140
Graph 9.1:	Comparison of Accrued Wages for Russian Workers by Selected Sectors from 1991 to 1996	168
Graph 10.1:	Mean CES-D Score by Income	205

Graph 10.2:	Mean CES-D Scale Score and Percentage of Respondents Above Depression Criterion by Size of Helping Network	206
Graph 10.3:	The Relationship Between Household Labor Groups and Symptoms of Depression as Mediated by the Size of the Personal Helping Networks	209
Graph 10.4:	The Relationship Between Gender and Symptoms of Depression as Mediated by the Size of Personal Helping Networks	210
Graph 10.5:	The Relationship Between Age and Symptoms of Depression as Mediated by the Size of Personal Helping Networks	211

List of Tables

Table 2.1:	Mean Percentage of Commodity Production by Large Enterprises, *Fermery* and Household Plots in Russia in 1996	22
Table 3.1:	Components of Rural Population Size Changes (in thousands)	46
Table 3.2:	Proportions of the Rural Population by Economic Regions of Russia from 1970 to 1997 (in percent)	47
Table 3.3:	Number and Size of Households in Rural Russia by Economic Regions (in 1994)	51
Table 3.4:	Age Structure of the Rural Population in Russia in 1997 (in percent)	54
Table 3.5:	The Structure of the Rural Population in Russia by Human Capital Age Groups and Economic Regions in 1997 (in percent)	55
Table 4.1:	Number of Households and Individuals, Sample Size and the Percentage of Households Interviewed in the Three Villages from 1995 to 1997	68
Table 4.2:	Cronbach Alpha Reliabilities for Various Life Domains and Life-as-a-Whole Indexes for 1995, 1996 and 1997	75
Table 5.1:	Demographic Types of Households in the Panel Study Sample from 1995 to 1997	80
Table 5.2:	Percentage of Households Which Decreased, Increased or Had No Change in Size from 1995 to 1997	83

Table 5.3:	Age Structure of All Household Members by Gender and Year	84
Table 5.4:	Marital Status of All Adults 18 Years of Age or Older in Respondents' Households in 1995, 1996 and 1997 (in percent)	85
Table 5.5:	Education Level of All Adults 18 Years of Age or Older in Respondents' Households in 1995 (in percent)	86
Table 5.6:	Average Number of Years of Education by Position for All Working Age Men and Women in Respondents' Households in 1995 (in percent)	87
Table 5.7:	Distribution of Household Labor in the Total Sample for Each Year of the Panel Study	88
Table 5.8:	Distribution of Household Labor by Demographic Type of Household in 1995 and 1997	89
Table 5.9:	Mean Distribution of Household Labor by Village For Each Year of the Panel Study	90
Table 6.1:	Respondents' Connections to Their Villages and the Local Area Through Birth and Kinship in 1995	95
Table 6.2:	Attendance at Family Festivals and Ceremonies	98
Table 6.3:	Attendance at Village Festivals and Ceremonies	98
Table 6.4	Mean Scores on the Index of Community Involvement	100
Table 6.5:	Mean Scores on the Index of Community Fit	101

Table 6.6:	Mean Social Network Indicators in 1995 and Changes from 1995 to 1997	103
Table 6.7:	Mean Community Attachment and Social Network Indicators by Level of Household Labor	105
Table 7.1:	Mean Size of Different Types of Land Use by Village and by Year	113
Table 7.2:	Mean Size of Rented Land by Household Labor	116
Table 7.3:	Mean Percentage of Rented Land by Level of Community Involvement	117
Table 7.4:	Rental of Land by Number of Non-Redundant Ties from 1995 to 1997	118
Table 7.5:	Percentage of Households Owning Agricultural Equipment and Tools in 1995, 1996 and 1997	119
Table 7.6:	Percentage of Households Using Selected Types of Agricultural Inputs in 1995, 1996 and 1997	120
Table 7.7:	Mean Number of Total Agricultural Tools (excluding tractors) by Village and Year	121
Table 7.8:	Mean Number of Total Agricultural Tools (excluding tractors) by Household Labor and Year	122
Table 7.9:	Mean Number of Total Agricultural Tools (excluding tractors) by Level of Community Involvement by Year	123
Table 7.10:	Percentage of Households Owning Different Amounts of Selected Types of Livestock	124
Table 7.11:	Mean Level of Household Ownership of Livestock in the Three Villages	125

Table 7.12:	Mean Weighted Number of Animals in Households by Village	126
Table 7.13:	Mean Weighted Number of Animals by Household Labor	127
Table 7.14:	Mean Weighted Number of Animals by Community Involvement	128
Table 8.1:	Mean Household Production of Meat and Milk (kilograms) in Three Russian Villages from 1995 to 1997	135
Table 8.2:	Mean Total Weighted Household Agricultural Production by Village	136
Table 8.3:	Mean Household Agricultural Production (kilograms) by Household Labor from 1995 to 1997	137
Table 8.4:	Mean Weighted Production (kilograms) by Number of Non-Redundant Social Network Ties	139
Table 8.5:	Mean Total Household Agricultural Production (kilograms) by Rental of Land	142
Table 8.6:	Mean Total Household Agricultural Production (kilograms) by Weighted Number of Animals	143
Table 8.7:	Standardized Regression Coefficients for Observed Exogenous Variables and Explained Variance (R^2) for Structural Model of Peasant Household Agricultural Production	146
Table 8.8:	Mean Total Weighted Household Sales of Different Commodities (kilograms) by Three Russian Villages from 1995 to 1997	149

Table 8.9:	Total Household Agricultural Sales (kilograms) by Village and Year	151
Table 8.10:	Mean Total Weighted Sales (kilograms) by Household Labor from 1995 to 1997	152
Table 8.11:	Mean Total Weighted Sales (kilograms) by Number of Non-Redundant Ties	153
Table 8.12:	Mean Total Weighted Sales (kilograms) by Level of Community Involvement	154
Table 8.13:	Mean Total Sales (kilograms) by Number of Animals	154
Table 8.14:	Standardized Regression Coefficients for Observed Exogenous Variables and Explained Variance (R^2) for Structural Model of Peasant Household Agricultural Production Sales	157
Table 9.1:	Employment Status of Working Age Adults in Three Russian Villages in 1995 and 1997	162
Table 9.2:	Types of Enterprises Where Working-Age Adults Are Employed in Three Russian Villages from 1995 to 1997	163
Table 9.3:	The Contribution of Different Sources of Income to Total Monthly Household Income in Three Russian Villages from 1995 to 1997 (in percent)	166
Table 9.4:	Income from Different Sources by Village in Three Russian Villages from 1995 to 1997	169
Table 9.5:	Income from Different Sources by Household Labor in Three Russian Villages from 1995 to 1997	173

Table 9.6:	Income from Different Sources by Level of Community Involvement in Three Russian Villages from 1995 to 1997	174
Table 9.7:	Income from Different Sources by Size of Non-Redundant Helping Networks in Three Russian Villages from 1995 to 1997	176
Table 9.8:	Standardized Regression Coefficients for Observed Exogenous Variables and Explained Variance (R^2) for Structural Model of Peasant Household Monetary Income	180
Table 9.9:	Differentiation of Income in the Three Villages by Year	182
Table 9.10:	Percentage of Households in Poverty in the Panel Study	183
Table 9.11:	Mean Number of Durable Goods Owned by Household Labor in Three Russian Villages from 1995 to 1997	184
Table 9.12:	Mean Number of Durable Goods Owned by Level of Community Involvement in Three Russian Villages from 1995 to 1997	186
Table 9.13:	Mean Number of Durable Goods Owned by Size of Non-Redundant Helping Networks in Three Russian Villages from 1995 to 1997	188
Table 10.1:	English and Russian Versions of the Modified Version of Center for Epidemiological Studies of Depression (CES-D) Scale	195

Table 10.2:	Descriptive Statistics of the Modified CES-D Scale in Rural Russia	196
Table 10.3:	Mean CES-D Score by Village and Year	198
Table 10.4:	Mean CES-D Score by Household Labor from 1995 to 1997	199
Table 10.5:	Mean CES-D Score by Year and Sense of Fit in the Village	199
Table 10.6:	Mean CES-D Score by Number of Non-Redundant Ties	200
Table 10.7:	Standardized Regression Coefficients for Observed Exogenous Variables and Explained Variance (R^2) for Structural Model of Depression	204
Table 10.8:	Direct, Indirect and Total Causal Effects of Personal Helping Networks on Symptoms of Depression	208
Table 11.1:	Mean Level of Satisfaction with Five Life Domains and Life in General by Year	217
Table 11.2:	Mean Level of Satisfaction with Five Life Domains and Life in General by Year and Village	219
Table 11.3:	Mean Level of Satisfaction with Four Life Domains and Life in General by Year and Household Labor	222
Table 11.4:	Mean Level of Satisfaction with Four Life Domains and Life in General by Year and Level of Community Involvement	225
Table 11.5:	Mean Level of Satisfaction with Four Life Domains and Life in General by Year and Sense of Fit in the Village	227

Table 11.6:	Mean Level of Satisfaction with Four Life Domains and Life in General by Year and Size of Helping Network	229
Table 11.7:	Standardized Regression Coefficients for Observed Exogenous Variables and Explained Variance (R^2) for Structural Model of Peasant Household Quality of Life	233

List of Contributors

David J. O'Brien is Professor of Rural Sociology, University of Missouri-Columbia. He is the principal investigator, American side, on the Russian Village Studies Projects. He has authored or co-authored books on Neighborhood Organization in the USA and Japanese American Ethnicity, as well as three books and numerous articles with Drs. Patsiorkovski and Dershem on rural life in Russia and the USA.

Valeri V. Patsiorkovski is a Laboratory Chief in the Institute for Socio-Economic Studies of Population, Russian Academy of Sciences, Moscow, Russia. He is the principal investigator, Russian side, on the Russian Village Studies Projects. He has been author, co-author or editor of 17 books and over 100 scientific articles on numerous issues related to services and the well-being of population in both urban and rural Russia and other Former Soviet Union countries. Dr Patsiorkovski has lectured extensively in the United States and Canada.

Larry D. Dershem is the Associate Director, Center on Rural Transitions, at the University of Missouri-Columbia. He has directed household level research in several post-Soviet states, including Russia, Georgia, Armenia and Azerbaijan. In addition, he is a consultant for several international aid and development organizations. Dr Dershem has published extensively on social network methodology in transitional economies.

Preface

This book reports on the findings of a three-year panel study of Russian rural household adaptation to a market economy. This study was funded by the National Science Foundation (SBR 9409936), in cooperation with the Russian Academy of Sciences. Additional funds were provided by the University of Missouri Research Board, University of Missouri Office of Development, University of Missouri Brown Fellowship and the International Research and Exchanges Board (IREX).

The conceptualization and methodology for the panel study are the result of a long-term collaboration between social scientists at the University of Missouri and the Institute for the Socio-Economic Studies of Population, Russian Academy of Sciences. David O'Brien and Valeri Patsiorkovski, the American and Russian Principal Investigators, first met in Moscow in December of 1989, only a month after the fall of the Berlin Wall. Their collaboration began on the basis of mutual interests in rural social services and the maintenance of viable rural communities in an increasingly global agricultural economy. During the early phases of this collaboration, invaluable historical perspectives and survey methodological expertise were contributed by Charles Timberlake and Larry Dershem, respectively. Sensitivity to ways in which Russian rural life is affected by the globalization of agriculture and food systems was contributed by Alessandro Bonanno. This early work produced Patsiorkovski and O'Brien (1996), *Research Methodology and Quality of Rural Life in Russia and the USA* and O'Brien and Patsiorkovski (1996), *Services & Quality of Life in Rural Villages in the Former Soviet Union*.

The books just mentioned, as well as Russian government statistics, show that household agriculture continues to be a significant part of Russian agriculture in the post-Soviet period. Moreover, the Russian-American research team's surveys in 1991 and 1993 show that the different capacities of households to produce and sell agricultural products is a critical source of differentiation in the material and psychological well being of rural families.

The collapse of the Soviet Union and the involvement of Larry Dershem in the research team as a full partner produced conditions that made possible the research project described in this book. First, the collapse of the command economy meant that rural Russian households, for the first time since the New Economic Program period in the 1920s, could actually expand their production and sales. Second, Larry Dershem's methodological expertise, especially in social network analysis, provided the impetus to measure how various types of household capital, including social capital, might affect economic adaptation, as well as mental health and subjective quality of life. A study that he spearheaded in 1993 provided an opportunity to refine the methodology that eventually became the basis for the panel research design.

Valentina Patsiorkovskaya managed the field staff that carried out the interviewing for the panel study, as well as the staff responsible for coding of completed surveys and the development of the SPSS statistical files. She has occupied this position in the Russian-American research team since 1991. Katherine Lyman provided critical statistical consultation on the panel design and analysis of data. Charles Timberlake continued to provide critical historical perspectives on the panel survey instrument and interpretation of the data. Others who made major contributions to the project include, R. Fleischmann, W. Handayani, J. Intajara, I. V. Korkhova, M. A. Kazlova, O. V. Krukhmaleva, I. N. Lisova, Qian Liu, O. V. Lylova, B. E. Murphy, A. I. Petrova, E. E. Ryzhkina and A. M. Semenova.

We would like to give special thanks to local government administrators who assisted with access to the study villages: Antonina Fedorenko (Latonovo), Aleksandr Putivtsev (Vengerovka) and Ludmila Shirokova and Tamara Ustiantseva (Sviattsovo). Fedor Redchenko provided access to the village of Mayaki in the earlier 1991 and 1993 studies.

We are also grateful for comments on various portions of the material included in this book and encouragement by the following scholars: Maria Amelina (World Bank), Cynthia Buckley (University of Texas-Austin), Nancy Popson (Kennan Institute), Richard Rose (University of Strathclyde), Blair Ruble (Kennan Institute) and Stephen Wegren (Southern Methodist University). A special thanks to Brigitte Holt and Heather Ramsay for completing the camera-ready copy of the manuscript.

Institutional support is vital to the success of a joint international research project. At the University of Missouri-Columbia, Brady Deaton and Mike Nolan had the foresight in the early 1990s to endorse this project.

We are especially thankful for the creative ways in which Mike Nolan provided resources to David O'Brien and Larry Dershem to travel to Russia before external funding was obtained and for salary support for Larry Dershem to work full-time on the project. Natalia Rimashevskaya, director of the Institute for Socio-Economic Studies of Population, Russian Academy of Sciences, has been extremely helpful throughout the project in providing us with scarce resources in Moscow and allowing Valeri Patsiorkovski to be away from his laboratory for extended periods of time.

Overall, this project could not have been accomplished without the warm reception and cooperation we received from residents in the study villages. They gladly took time from their busy schedules to accommodate the research team's questions. Finally, although numerous people and institutions contributed to the creation of this book any shortcomings, mistakes and errors remain solely those of the authors.

1 Changes in Russian Agriculture: New Institutional Structures in the Countryside

David J. O'Brien, Valeri V. Patsiorkovski and Larry D. Dershem

Introduction

Since the collapse of the Soviet Union, most Western observers have agreed that major structural changes must be made in the organization of Russian agriculture if it is going to compete in a global economy. The question remains, how can such a restructuring occur within the constraints of the historical and contemporary Russian economic, political and social situations?

In order to truly transform rural institutions to support a free-market agriculture, it will be necessary to provide the wherewithal to develop the core social organizational unit in the Russian countryside, the *krest'ianskoe khoziastvo* or peasant household. *The peasant household is more than just an agricultural production system.* Its human and social capital have the potential to play a central role in the development of the social and cultural infrastructure that is necessary to build the "civic society" that will support both a market farm economy and a democratic society. In this respect, our view of the economic development of Russian agriculture is closest to the traditional liberal-democratic view expressed by Tocqueville, that the economic institutions of a democratic society ultimately rest on the development of social institutions within which economic relations are embedded. The more recent expression of this view is found in Putnam's (1993) work on "social capital," and North's (1990) work on institutional economics which suggests that countries that are successfully competing internationally have developed institutions of cooperation and trust that give their citizens an advantage over citizens in other nations.

Ultimately, the restructuring of Russian agriculture will require a major overhaul of Russian land law and the growth of private farming. The difficulty, however, is to identify the specific causal mechanisms through which this transformation might take place. The prevailing view, among most Russian reformers and Western observers, has focused on legislative and legal changes; specifically, attempting to create a land market and officially registering a new class of private farmers. Western observers have, by and large, judged these legislative and legal efforts as failures. As a result, there has been an image of very limited change in the Russian countryside.

Overlooked in the midst of the attention given to land reform and the creation of a new farmer class, however, are incremental but fundamentally crucial changes that are occurring in the *krest'ianskoe khoziastva*. Because there has been virtually no systematic empirical analysis of how peasant households are adapting to the new market economy, these changes are barely known in Moscow, let alone in the West. *The primary purpose of this book is to analyze the changes in the nature of peasant households and how these changes will impact on the eventual development of a sustainable agriculture in Russia and its integration into the global economy.* In our view, the development of peasant household production is the best hope for the overall development of Russian agriculture because it rests on a solid base of human and social capital (O'Brien et al 1998a, 1997, 1996a, 1996b, 1993; Patsiorkovski and O'Brien 1996). Rather than trying to import human and social capital from urban areas, we will argue that a simpler and more efficient way to develop Russian agriculture is to use existing strengths in the countryside that are found at the household level.

In order to understand the changing role of the peasant household in Russian agriculture, it is necessary to illustrate the structure of institutions in rural Russia during the Soviet period and during the contemporary transition to a market economy. Figures 1.1 and 1.2 below illustrate these institutional changes.

Agriculture and Rural Life in the Soviet Period

The relationships between the various institutional elements in the Russian village during the Soviet period are shown in Figure 1.1. This period began with forced collectivization in 1929 and continued until the breakup of the Soviet Union in 1991.

The core institution in the Soviet agricultural system was the large state enterprise, the *kolkhoz* (collective farm) or *sovkhoz* (state farm). These large enterprises were designed as agricultural factories that received production quotas (*obiazatel'naia postavka*) from government officials. Being outside of a market economy, the managers of these large enterprises were primarily concerned with staying in the good graces of officials at higher levels. Questions concerning where grain or meat was to be processed and sold were handled administratively at higher levels, beginning in Moscow, and then transmitted by decree to the regions, provinces and, eventually, to the large enterprises themselves.

Moreover, the large enterprises also were entrusted with a very diffuse set of responsibilities. These included: providing consumer goods, through village shops; social services, such as health and education; material infrastructure, such as roads and utilities; land management; and employment of all able-bodied residents in rural villages. Because of technological advances, Soviet agriculture became more capital intensive over time and thus became less dependent on local village labor. By the end of the Soviet period there was a substantial surplus of labor attached to the *kolkhozy* and *sovkhozy*.

In the Soviet system, the local village and the households within it, were expected to do the bidding of the large enterprises. Local villages had two main functions. The formal function provided by the local government, the *sel'soviet,* was to keep records of the demographic structure of the households in the villages, such as births, deaths, and marriages, household plots, and housing. In effect, this served the needs of the large enterprise by keeping track of the condition of its labor force. The informal social function of the village was to provide individuals and households with an opportunity for informal community interaction. In most instances, the village pre-dated the Soviet collectivization and provided individuals and families with a psychological and spiritual connection with their ancestors and history.

In the Soviet system households were expected to serve five basic functions. First, they were expected to provide labor for the *kolkhoz* or *sovkhoz*. Second, households were expected to socialize children and maintain informal constraints on adults to support the goals of the large enterprise and thereby support the goals of the Soviet State.

Third, households were seen as important consumers within the Soviet command economic system. Their preferences often were not treated with a great deal of respect, especially in comparison to their more politically powerful urban-industrial counterparts. Fourth, households were involved in and maintained the village community. This was encouraged by the large enterprises because it was necessary to sustain the communal basis of their labor force. The nature of that participation, however, was severely restricted by Soviet fears about competition from non-governmental associations.

Figure 1.1: Institutional Structure of Rural Life in Soviet Russia

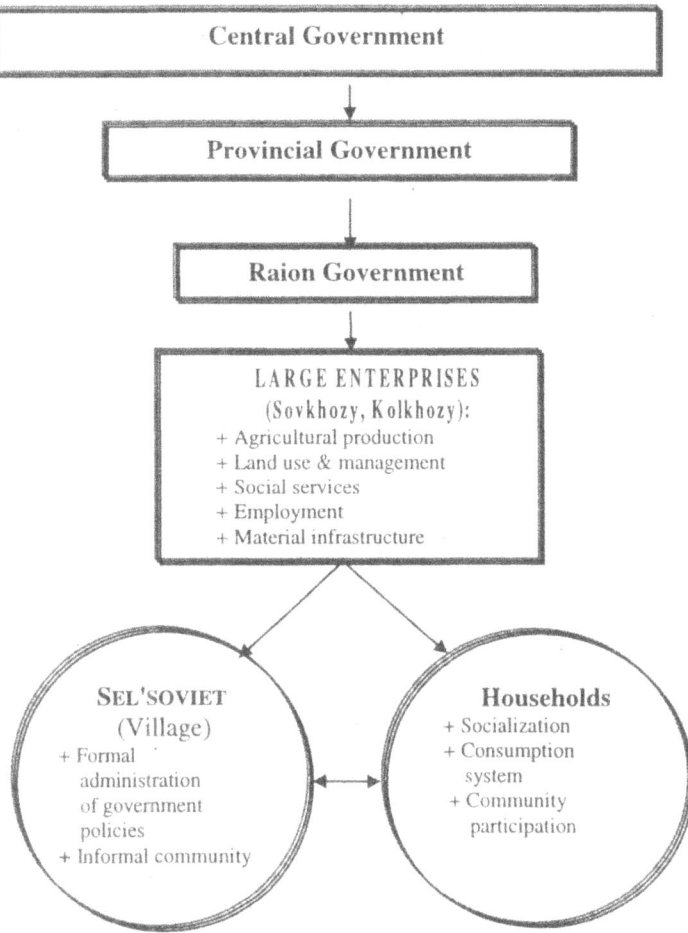

Finally, households were allowed to keep small plots of land around their house, normally not more than one-third of a hectare, on which they were able to grow food and raise animals for their personal consumption. This provided a positive incentive for households to remain in the countryside, as well as providing a safety valve for the state when production quotas were not met.

There were, however, constant struggles between households and the government on the priority of these functions. The government placed primary emphasis on the labor function of households for the *kolkhozy* and *sovkhozy*. Households placed greater emphasis on socialization, consumption and occasional sales. These different priorities created considerable tensions between rural households and the government during the entire Soviet period.

Agriculture and Rural life in the Post-Soviet Period

Reports on attempts to restructure Russian agriculture, from both Russian and Western sources, usually give the impression that there has not been very much real change since the collapse of the Soviet Union. Since 1990, the executive branch of the Russian federal government (President Yeltsin) has issued a series of decrees that outline a program for selling land to private persons. The legislative branch, the Duma, led by the Communist Party, however, has balked at these decrees and has refused to pass land reform legislation that would be signed by the executive branch. This struggle has resulted in a stalemate that has left people at the local level in a state of uncertainty regarding ownership of land.

Discussions of land reform have centered on breaking up the large enterprises, the *kolkhozy* and *sovkhozy*. These efforts, however, have not been very successful. The large enterprises continue to exist, although they have been officially reorganized, usually into "joint stock companies of the closed type" (*tovarishchestvo s ogranichennoi otvetstvennostiu*, or *TOO*s). The *TOO*s only permit membership and share ownership from collective farm members and individual members do not have any real input into decision-making, which is left to the collective farm chairman. In short, the *TOO*s are really not that different from the *kolkhozy* they were supposed to replace.

Despite the aforementioned, there have been some important incremental changes in the institutional relationships in agriculture and rural life in the post-Soviet period. These are shown in Figure 1.2.

Although the large enterprises continue to exist, their relationships to the Russian government, the villages, and the households have changed. Government subsidies to the large enterprises have declined continuously since the breakup of the Soviet Union. Consequently, a large proportion, 82 percent in 1997 (Semenov 1998), of the enterprises simply have gone bankrupt. Those that remain must compete in the marketplace. In 1994, the

Changes in Russian Agriculture 7

government decreed that the large enterprises stop providing social services, although many continue to do so. This has meant that the large enterprises have divested their responsibilities for non-agricultural activities, such as support for social services, to the local village council (formerly sel'soviet) or, increasingly, to the private sector. As was shown in Figure 1.1, the large enterprise was the only mediating institution between other rural institutions and the government in the Soviet period. Now, as shown in Figure 1.2, the local village council mediates between all rural institutions and the central and provincial governments. The traditional large enterprises are one of several enterprises in rural areas.

Figure 1.2: Institutional Structure of Rural Life in Contemporary Russia

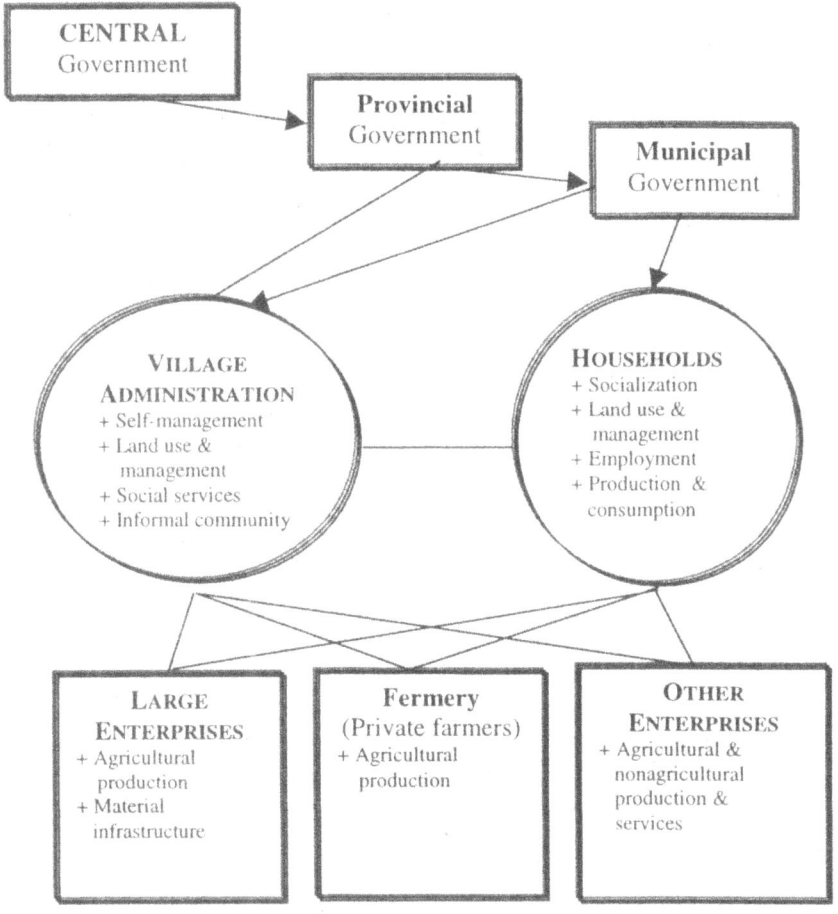

Local government, which is more representative of local village interests than the old s*el'soviet*, now is the official link between the national and regional governments and the village. Because the federal government has been in an almost constant fiscal crisis since 1991, it has not been able to provide much support for local services and thus rural villages have experienced an overall decline in social service support. Yet, the informal relationships between individuals and households in the village community continue to exist, although the strength of these relationships has diminished (see Chapter 6).

One of the new types of enterprises in rural areas is the new private farmers (*fermery*). *Fermery* are registered as individuals, not as households. The numbers and political influence of this group, however, has remained small. From 1993 to 1994, the number of persons registered as *fermery* increased from 182,800 to 270,000, but the size of this group increased only slightly in 1995 (279,200) and in 1996 (280,100) and actually decreased in 1998 (274,000). The total amount of land held by officially registered *fermery* increased from 7.8 million hectares in 1993 to 12.2 million hectares in 1997 (about six percent of the arable land in Russia). The average amount of land farmed by persons in this status has changed only slightly, from 43 hectares in 1993 to 44 hectares in 1997 (Russia in Figures 1997:298; *Agranaia Reforma Rossi* 1998). Approximately two-thirds of the newly registered *fermery* in 1991 were urban dwellers who typically had very little direct experience with agriculture. Because of their relatively small impact on total agricultural production, the *fermery* remain politically weak (Wegren 1998:10, 19-22), which prevents them from exerting a significant influence on national or *oblast* agricultural policies.

The inability to induce families to become private farmers is based in part on the failure of the federal government to resolve the land tenure problem. Ordinary rural Russians, who are rational economic actors, have very reasonable fears that if they buy land they may end up losing it and/or paying some type of penalty for it at some later point in time. There are, however, four other obstacles faced by persons in this new status. First, the general economic uncertainty in Russia reinforces a natural caution that peasant households have with regard to risk. Second, there is a lack of suitable infrastructure for small-scale processing plants and lack of support for small business services. Third, high tax rates create further disincentives to officially register as *fermery*. Finally, families that become disconnected from the large enterprises no longer can receive valuable benefits, such as discounted meals in the *TOO* cafeteria or help in getting supplies and inputs for agricultural operations. As a result, most peasant households operate in an informal economy in which their income is shielded from the official tax collection agencies.

The functions of the household in rural areas have changed from the Soviet period. First, recent laws on privatization have made households the legal owners of land that was formerly owned by *kolkhozy* and *sovkhozy*. At this time, the vast majority of households do not have title or personal use of that land. Nevertheless, all large enterprises must rent their land from rural households. Second, with the increase in other types of enterprises, household members can choose where they would like to work. Third, households continue to be consumers, but they have become significant agricultural producers in recent years. All of these changes are shown in Figure 1.2 as interdependent relationships (two-headed arrows).

Although they have not received much legal protection, land, or credit from the federal government, peasant households have made substantial changes during the past few years. Households continue to be a source of labor for the large enterprises, and an important social organizational unit for socialization, consumption and community participation, but critical changes have occurred in their agricultural production. The elimination of restrictions on the ability of households to sell products in the marketplace has provided a powerful stimulus for the development of household agricultural production.

Household Survival Strategies

The crisis in the Russian economy, including the agricultural sector, caused hyperinflation during the first few years of the post-Soviet period. This meant that many Russians lost their entire life savings and at times had to go without many of the necessities of life. In spite of enormous difficulties at the macro-economic level, however, Russian households have made remarkable adjustments in their own *micro-economies*. It is at this level where some of the most creative adaptations to the market economy have occurred.

Nowhere has the adaptation of households been more remarkable than in the Russian countryside. Despite the failure of the central government to resolve the problem of land ownership and the absence of any real restructuring of the large collective farms, the average amount of agricultural products sold by peasant households, *krest'ianskie khoziastva*, has increased markedly since the end of the Soviet period. From 1994 to 1996, for example, the proportion of the total amount of meat produced in Russian agriculture accounted for by these small households increased from 43.2 to 51.5 percent. The proportion of milk products, including milk, butter, sour cream and cottage cheese, produced by these households increased from 38.7 to 45.3 percent during the same time period (Russia in Figures 1997:296). This was accomplished by making major adjustments in household practices, often introducing value-added processing to meat and dairy commodities they had produced. Many households have developed complex marketing strategies with members of their extended families and other households to sell products in hotels and restaurants in urban centers, as well as in urban farmers' markets.

Although official figures showed their income was dropping, because of the lack of wage payments from the reorganized collective farms, many rural households have been able to purchase new durable goods during the past few years. From 1995 to 1997, the percentage of households in our three study villages who owned automobiles increased from 16.4 to 21.8 percent, which is an increase of almost one-third in a short three year period. The percentage of households with telephones went from 14.7 to 18.4 percent and the percentage of VCRs went from 8.4 to 19.0 during the same period. These data indicate what Erlanger observed,

> 'I find ordinary Russians adjusted to the new world and the new economy much faster than any theorist expected them to. They found ways to make money, sometimes not very lovely ways, sometimes ways in which they felt degraded, but ways that kept their families going' (Shearer and Starr 1996:39).

The success of individual households in adapting to the market economy has varied considerably, although the gap between the well off and the not so well off has not been nearly as great as it is in large metropolitan areas, like Moscow and St. Petersburg. Although some individuals have made substantial economic gains in the post-Soviet period, one is not likely to see "New Russian" tycoons driving fancy automobiles in rural Russian villages. Moreover, rural family and neighbor support networks do mitigate some of the harsher consequences of a market economy for those who are dependent. In this respect, the more vulnerable portion of the rural population would seem to have a significant advantage over its urban counterparts. A visitor to the Russian village will not encounter the beggars or homeless people who have become commonplace in large cities.

At the same time, however, households vary considerably in their capacity to take advantage of an economic niche in the new economy, as we will show. Those households with the highest amount of labor, as indicated by the number of working age adults, produce almost four times more agricultural commodities than their neighbors with the lowest amount of household labor. Households with higher amounts of *social capital*, as indicated by having more persons in their helping networks and being more involved in the social life of their villages, have additional advantages over their neighbors. In turn, households with more human and social capital are able to increase their advantages by gaining access to various types of physical capital. Since agricultural sales account for roughly one-third to one-half of household income, on average, these advantaged households have significantly higher purchasing power than do other households.

The different degrees of success of Russian rural households in making adjustments in their micro-economies can be viewed in two ways. One view is that this is merely a temporary adjustment until more substantial macro-level changes are made to truly reform the Russian economy. From this perspective, although the creative adaptability of Russian households is admirable, their "making do" with present circumstances does not really produce lasting reforms.

A second view, and the one we share, is that although macro-economic adjustments are crucial for reform of the Russian economy and Russian society, the micro-level adjustments of households are more than temporarily "making do" with a transitional situation. Rather, *these micro-level adjustments are based on a fundamental reorganization of peasant household production and sales, and rural institutions, that will provide the foundation for the future development of the agrarian sector in Russia. In turn, these adjustments mean that in order to understand and assist in the process of reforming Russian agriculture it will be necessary to understand the role of human, social, and physical capital within the peasant household.*

The basic social organizational unit of Russian agriculture in the pre-Soviet period was the *krest'ianskoe khoziastvo*. Peasant households produced agricultural commodities using their own labor, largely hand labor, and developed informal production, processing and marketing networks with their kin and neighbors. The Soviet period changed this entirely (as shown in Figure 1.1 above), by forcing peasant households to become suppliers of labor for the state sanctioned large agricultural enterprises, the *kolkhozy* and *sovkhozy*. The Soviets did permit peasant households to produce commodities for consumption and for sale on small household plots (usually one-third of a hectare or less). Nevertheless, the state monopolization of processing facilities and markets, and, at times outright repressive measures, created disincentives for households to improve their human or social capital.

Moreover, the large enterprises also were entrusted with a very diffuse set of responsibilities. These included: providing consumer goods, through village shops; social services, such as health and education; material infrastructure, such as roads and utilities; land management; and employment of all able-bodied residents in rural villages. Because of technological advances, Soviet agriculture became more capital intensive over time and thus became less dependent on local village labor. By the end of the Soviet period there was a substantial surplus of labor attached to the *kolkhozy* and *sovkhozy*.

In the post-Soviet period, state monopolization of markets has weakened. Outright prohibitions against market-place activities have virtually ended, and even though the tax structure would be considered unduly burdensome by Western standards, peasant households have found ways to get around these constraints and develop an informal economy. These changes, coupled with the inability of the large enterprises to provide wages in cash, have produced strong incentives for peasant household members to see themselves as members of household economic enterprises rather than as *kokhozniky*.

The persistence of the large enterprises, the *kolkhozy* and their re-organized forms, the *TOOs*, and the apparent inability of the Russian central government to produce effective land-reform legislation (The Economist 1998) may obscure these profound changes in Russian agrarian life. Moreover, these household enterprises are quite small by Western standards. Nevertheless, their recent growth represents, in a nascent form, a social organizational structure that, in our view, eventually will produce much larger farms that can become the basis for a sustainable Russian agriculture.

This view of the peasant household as a viable agricultural production unit does not mean, of course, that all rural households contribute in significant ways to the Russian agricultural economy today, or that all current producers will be contributors to that economy in the future. Extrapolations from our findings suggest that at least 50 percent, or roughly

7 million, of the 14 million rural households in Russia today make some type of contribution to agricultural sales. A conservative estimate, based on our findings (see Chapter 3), would be that roughly half, or 3.5 million, of these households, will be able to expand their production capacity sufficiently to continue to be viable farm households as the Russian agricultural economy evolves.

The key research task, and the central focus of our panel study upon which this book is based, is to identify specifically how the different levels of capital (human, social, and physical) in the peasant household are transformed into varying degrees of success in adapting to a newly emerging agrarian system in Russia. The completion of this task will have two benefits for public policy formation, both for Russians and for international organizations that have an interest in the development of the Russian agrarian sector. The first of these is to better understand how to improve the efficiency of Russian agriculture by stimulating the growth of household capital. The second benefit will be to identify ways in which selective intervention by government, non-governmental organizations, and other interested parties might assist those households with limited human or social capital that are having difficulty creating household enterprises.

Our theoretical approach to the Russian peasant household, which focuses on variations in levels of household capital, will be developed in Chapter 2. In Chapter 3 we will examine macro-level data on the size and structure of rural households throughout Russia, thereby placing our micro-level household and village study within a broader context that recognizes the enormous diversity of the Russian countryside. In Chapter 4 we will outline the background and rationale for the three-year panel study (1995 to 1997) in three Russian villages. Chapters 5, 6, and 7 present empirical findings on human, social, and physical capital, from the panel study.

Chapters 8 and 9 present findings on the impact of different levels of all types of household capital on agricultural production and sales, and total family income (both monetary and nonmonetary). Chapter 10 examines the relationship between household capital and stress, as it is reflected in symptoms of depression. Chapter 11 presents findings from the panel study on the relationships between household capital and subjective quality of life. In Chapter 12 we will make recommendations on ways to support sustainable households and communities in an era of globalization in rural Russia.

2 Peasant Households and the Transition to a Market Economy and Democracy

David J. O'Brien, Valeri V. Patsiorkovski and Larry D. Dershem

Introduction

Scholarly approaches to the transition of Russian agriculture to a market economy tend to focus on privatization of land and the development of new efficient agricultural enterprises and a class of private farmers. Because results in both of these areas have been disappointing (Foundation for Support of Agrarian Reform and Rural Development 1998; IFC 1996), there has been a sense that very little structural change has taken place in rural Russia since the collapse of the Soviet Union.

We will offer an alternative perspective, proposing that, although incremental in nature, important structural changes are taking place which have important consequences for the future of Russian agriculture and the development of new forms of social and economic stratification in the Russian countryside. Our argument rests on the examination of the traditional basic social organizational unit of Russian agriculture, *krest'ianskie khoziaistvo*, or peasant household. Overall, peasant households have been adjusting to the impasse over land reform and the creation of a new class of farmers by developing their own *human and social capital* at the household and village levels. There have been, however, substantial variations in the ability of households to improve their human and social capital, which has led to increasing differentiation in the capacity of households to produce agricultural products for market. This, in turn, has led to increasing differentiation in the quality of life of Russian rural households, reflected in differential access to durable goods, social service needs, and levels of stress.

16 Household Capital and the Agrarian Problem in Russia

There are two further implications of these differences in the capacity of rural households to compete in an emerging market economy. The first is that they will create differences in the levels of citizen support for market reforms and democratic institutions. The second is that local *Oblasts* (provinces) and villages will become differentiated according to their ability to support the growth of human and social capital in peasant households.

The Land Reform Stalemate

The problems of Soviet agriculture have been well documented. These include enormous waste and inefficiency, lack of adequate storage and transportation facilities, cultivation of crops on land for which they were ill-suited, environmental degradation, and the inability of large enterprises to respond to rapid shifts in consumer preferences. As a result of these conditions, Soviet agriculture increasingly became dependent on large subsidies from the central government, which drained much needed capital from other parts of the Russian economy (Karcz 1971; Yaney 1971). Given this history, it is not surprising that most Western, as well as many Russian, observers saw the solution to the problems of Russian agriculture as simply breaking up the large Soviet-style enterprises, the *kolkhozy* and *sovkhozy*, and creating a whole new class of farmers who were entirely independent from them (Prosterman and Hanstad 1993; Smith 1990:215,229; Van Atta 1993).

Efforts to break up the large agricultural enterprises, however, have had very limited success so far. In 1990, a land reform law was passed in the Russian Federation, and since the collapse of the Soviet Union a number of Presidential decrees have been issued which ostensibly were designed to promote land reform. The conservative, Communist-dominated, Duma, which, under the Russian Constitution (both pre- and post-1993), has the legitimate authority to pass land reform legislation, however, has continued to resist passage of legislation which would create a true land market in Russia (see Appendix 1). There continues to be a stalemate between the executive and legislative branches of the Russian central government over land reform legislation (O'Brien et al. 1998a; Wegren 1998). As a result, there has been very little change in land ownership since the collapse of the Soviet Union. Wegren observes that:

Today [1997], the 'land market' in Russia is primarily characterized by the exchange by individuals of small land plots, often less than one hectare in size. Enterprises [i. e. the *kolkhozy* and *sovkhozy*], for a variety of economic reasons, do not appear to be significantly active in the purchase of rural land, although there have been reports of banks and certain businesses purchasing whole farms in order to supply their workers with food. In general, more than three years after the legislative basis for a land market was established, the land market is limited, with less than 0.3 percent of eligible land involved in a transaction (1998:18-19).

Conservative politicians' (i. e., Communist party) resistance to land reform is tied to their power base which includes managers and workers in large-scale Soviet era enterprises, like the *kolkhoz* and *sovkhoz* farms, which oftentimes are not competitive in a true market economy and thus they have resisted privatization. But, it is also important to point out that there is widespread popular support for the conservative position in rural areas, which was reflected in support for Communist party candidates in recent elections. In the final round of the 1996 Presidential elections, for example, the Communist candidate Ziuganov did very well in the agricultural regions in the so-called "red belt" area (Central Black-earth Region) of Russia. In Belgorod *Oblast* (province), for example, he received 58.2 percent of the vote, while Yeltsin received 36.6 percent. This compares with Moscow *Oblast* (province) where Ziuganov received 29.4 percent of the vote and Yeltsin received 64.8 percent (Center for Russian Studies 1998).

Voting for the Communists is due to a genuine concern by rural Russians that the reformers have not really understood the consequences of privatization of agriculture for ordinary households (Hough 1994; Hough et al. 1996:4; Wegren 1998:22). Support for the popular view that the government is insensitive to the needs of rural citizens is found in Prime Minister Viktor Chernomyrdin's resolution No. 791 of July 6, 1994, in which there was only one sentence in twenty-three pages which addressed the problems of the rural household (APK 1994:3-25).

Most important, the Soviet agricultural model created a system whereby rural households became almost totally dependent on a service delivery structure that was linked to *sovkhoz* and *kolkhoz* farms. The reformers in the Russian government have not addressed the issue of replacing services that traditionally have been delivered, in part or in whole by the *kolkhozy* and *sovkhozy*. These services have included health care, education, utilities, groceries and dry goods, as well as assistance with private plots such as cultivation, access to inputs (e.g., fertilizer), and transportation to markets. At the same time, local government, the village council (formerly *sel'soviet*) which was very weak during the Soviet period, restricting its activities largely to recording births, deaths, and other types of demographic data on households, has not been able to gain resources to provide the full range of services needed in rural areas. Thus, rural residents have very reasonable fears about where they would be able to receive basic services if they become totally separated from the large agricultural enterprises (O'Brien et al. 1993).

The inability of the Russian central government to deal with the land issue, in a way which would provide some confidence to rural residents that they would be able to retain control over land if they did purchase it, and its' failure to provide reasonable access to services apart from the large enterprises have created disincentives for Russian households to formally register as *fermery*. It is these disincentives, along with a high tax rate on private farmers, rather than some supposed cultural resistance to private farming, which explains the resistance of Russian households to operating independently. From January 1991 to January 1996, the amount of land held by officially registered *fermery* increased from 181,084 to 12 million hectares; the latter being equivalent to about six percent of agricultural land in the country (Popov 1996).

As noted in the last chapter, since 1994, the creation of new *fermery* has slowed down considerably and the number of farm closures has increased Most telling is that approximately two-thirds of the newly registered *fermery* in 1991 were urban dwellers who typically had very little direct experience with agriculture. Moreover, because of their relatively small impact on total agricultural production, the *fermery* remain politically weak (Wegren 1998:10, 19-22), which prevents them from exerting a significant influence on national or *oblast* agricultural policies. Ordinary rural Russians, who are rational economic actors, have very reasonable fears that if they take land they may end up losing it and/or paying some type of penalty at some later point in time, or even lose their place to live (Wegren and Belen'kiy 1998).

Agricultural Producers in Russia Today

In Russia today, there are three main types of agricultural producers: the newly created *fermery*, the reorganized *kolkhozy* and *sovkhozy*, and *krest'ianskie khoziaistva* (see Figure 1.1 and Figure 1.2 in Chapter 1). As noted above, the *fermery* remain a small and not very influential portion of Russian agricultural producers. The other two types of producers, however, retain very influential, but quite different positions in overall agricultural production in Russia.

A restructuring of the collective farms began in 1992. Members of each *kolkhoz* were given several options. The first was for the farm to disband, distributing all land and property to farm members. This option was obligatory for chronically unprofitable farms. A second option was to remain a *kolkhoz*. By the end of the first phase of farm reorganization, 1992-1994, about one-third of collective farms chose to retain this status. The third option was to reorganize by creating a partnership of "limited responsibility," of the "open type." Persons outside the *kolkhoz* as well as present members of the *kolkhoz*, could purchase stock in the re-organized *kolkhoz* using their land and/or property shares. Remuneration and responsibility would be based on the profits of the re-organized *kolkhoz* and an individual's shares in that company.

The fourth option was to create a joint stock company of the "closed type" (*tovarishchestvo s ogranichennoi otvetstvennostiu*, or *TOO*). The *TOO*s only permit membership and share ownership from farm members. This variant has been the most popular of the reorganization options, chosen by about 47 percent of former *sovkhoz* and *kolkhoz* farms. Individual members receive shares, which in the *chernozem* area of south Russia are approximately four to five hectares per farm member. This allocated land cannot be sold to anyone. This does not pertain to the private plots and does not preclude informal rent relationships between village residents and *fermery*. If, for some reason, a member left the *TOO* he or she would be compensated for his or her share in cash. The remuneration for each household would be dependent on the profits of the *TOO* and the number of members of that household who were working for it. Individual members must work for the *TOO*, so that they are in virtually the same position as *kolkhozniky* working on *kolkhoz* farms. By and large, these reorganized collective and state farms remain very large, capital intensive, operations which concentrate on production of grains, meat or dairy products.

The third group of food producers are comprised of *krest'ianskie khoziaistva*. The term "peasant households" is a literal translation from Russian and is meant to convey a specific meaning of the household as a "basic unit of production and social livelihood" (Shanin 1982).

By world standards, persons living in rural Russian households have relatively high levels of education, they have been exposed to urban influences, and they have had regular access to television, radio, and newspapers from Moscow and from foreign countries (Humphrey 1988:54). Russian rural households, therefore, are not illiterate, nor are they isolated, and thus they are not like the persons who are typically described as possessing a "peasant culture" (Patsiorkovski and O'Brien 1996). Nevertheless, persons living in rural households in Russia refer to themselves by the term "peasant household" (O'Brien et al. 1996b).

Graph 2.1 shows the proportion of the different types of agricultural producers in Russia today, the proportion of land each holds, and the proportion of overall agricultural production contributed by each.

The large farms, the *kolkhozy, sovkhozy* and *TOO*s, are a relatively small proportion of the total number of agricultural producers, but they control close to 90 percent of all of the land under cultivation in Russia. In addition, they account for well over half of all agricultural production. The *fermery* are a very small proportion of total agricultural producers, control a very small amount of land and contribute very little to overall agricultural output in the country. Peasant household private plots, which average approximately one-third hectare in size, cultivate a very small proportion of the total agricultural land, a little more than 4 percent, but they are, by far, the largest number of agricultural producers in Russia. Most important, they contribute a significant share of overall agricultural production in Russia, approximately 46 percent.

Graph 2.1: Percentage Representation of Agricultural Producers, Land Used, and Mean Percentage of Commodities Produced by Large Enterprises, *Fermery* and Household Plots in Russia (1996)*

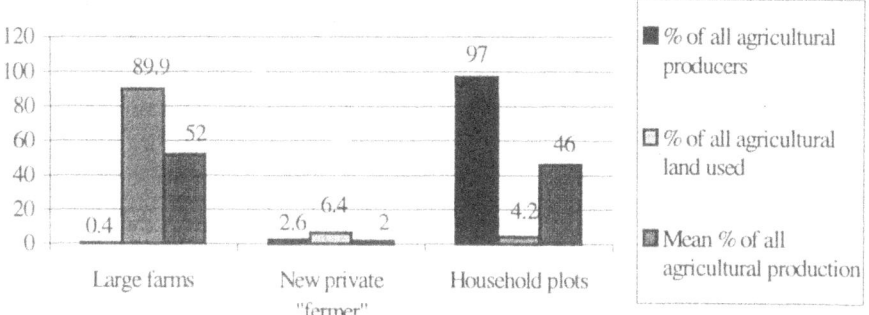

*Source: Ekonomicheskaia Gazeta (1994: 18); Goskomstat (1997: 294, 295).

The large farms and the peasant household plots are involved in different types of agricultural production, as shown in Table 2.1. Essentially, the large farms almost completely dominate production of capital-intensive large-scale grain production that requires very expensive equipment, large tracts of lands, and expensive inputs, like chemical fertilizers. They have substantially less involvement in more labor-intensive production activities, such as those involved in the production of meat and dairy products. The peasant household plots, on the other hand, contribute a significant share of production in those areas that are more labor intensive, such as meat, vegetables and potatoes. Most significant is that the trend of the large farms specializing in capital intensive production while the peasant households specialize in labor intensive activities has become more pronounced. The share of production of meat, potatoes and vegetables contributed by peasant household plots has actually increased in recent years.

Table 2.1: Mean Percentage of Commodity Production by Large Enterprises, *Fermery* and Household Plots in Russia in 1996*

Agricultural Producers	Commodities						
	Grain	Meat	Milk	Eggs	Potatoes	Sunflowers	Vegetables
Large farms	94.6	46.9	53.2	68.3	8.9	87.0	22.1
New private "*fermer*"	4.6	1.6	1.5	0.4	0.9	11.4	1.1
Household plots	0.8	51.5	45.3	31.3	90.2	1.6	76.8
Totals	100.0	100.0	100.0	100.0	100.0	100.0	100.0

* Source: Goskomstat (1997: 296).

The specialization of large enterprises in capital-intensive grain production may also be linked to the increase in imports of meat and dairy products from other nations, which have reduced their competitive advantage in those areas. However, peasant households are better able to fit a niche of labor intensive production, such as meat and vegetables. Their products typically are sold in the local *rynok* or farmer's market.

Nevertheless, individual households have had varying degrees of success in adapting to social change. Even prior to the formal collapse of the Soviet Union, there were indications that some households and villages were doing better than others in the nascent market economy. A 1991 survey found that households differed markedly in the amount of products they sold which were produced on their private plots, the extent to which they supported plans to privatize services, and the extent to which they were stressed by the economic crisis facing the country at that time. Households differed not only in their ability to participate in an emerging market economy, but their degree of participation was strongly associated with their support for economic reforms in general as well as their support for democratic reforms. *Most important, these differences between households were associated with differences in household capital, that is their ability to obtain human [household labor] and social capital [social networks and community attachment]* (O'Brien et al. 1993; O'Brien et al. 1998a; O'Brien et al. 1998b: 117-130,157-160,199-220; Patsiorkovski and O'Brien 1996: 140-147, 158-179).

Household Labor and the Differentiation of Peasant Household Economies

There are three characteristics of contemporary peasant household production in Russia, which are most relevant to understanding the emerging differentiation between households in their ability to take advantage of new opportunities for the production and sale of agricultural products. The first is that *peasant household production on small plots is very labor intensive*, by and large, involving the production, processing and sale of meat and dairy products. Feeding and care of animals and processing meat and dairy products requires a steady commitment of labor. The ability of a household to increase its number of animals, therefore, is directly correlated with its ability to obtain this labor from either within or from outside of the household.

The second distinguishing characteristic of peasant household production is that it responds to a very *narrow niche of opportunity, usually in local markets*. This niche, which is sometimes referred to as "truck farming" in the United States, includes farmers' markets as well as restaurants or hotels in a regional center which are located within driving distance of the village where the peasant household lives. Responding to this changing niche of opportunity requires that the household be able to produce, process and move products very quickly to market. The economic success of the household is especially dependent on its ability to process "value added" products, such as hams, sausages, sour cream and cottage cheese and to move them to urban markets as the demand occurs. Again, this calls for the ability to draw upon a dependable supply of labor.

Third, the household needs to be able to access inputs (e.g., fertilizers, pesticides, etc.), cultivation equipment, know about emerging niche markets, and find the means to transport its goods to market. Typically, the household does not have the resources on its own to accomplish these tasks and needs help from other households, as in a cooperative marketing venture, or from the *TOO* or *kolkhoz*, as in the case of accessing expensive cultivation equipment or trucks to transport products to market (O'Brien et al. 1993).

Our central thesis is that the ability of a household to meet the three types of needs just described will be proportional to its ability to obtain four types of capital; (1) household labor; (2) social exchange helping networks; (3) community attachment; and (4) physical capital.

Household Labor and Peasant Household Production and Sales

Our first hypothesis can be stated as follows:

H-1: *Households with more household labor, manifested in a greater number of working age adults in the household, will have a distinct advantage over other households in (a) agricultural production and (b) sales.*

Comparative studies of economic performance usually focus on *human capital* as a crucial explanatory variable. The term human capital typically is used in these analyses to refer to the amount and quality of education possessed by a household, community or nation. By world standards, human capital is quite high in rural Russia, as it is throughout Russian society. During the Soviet period, a considerable amount of resources were expended to meet the goal of educating children. In spite of the fact that rural areas were slighted in this regard, the educational levels of Russian rural households are considerable higher than those found in most rural regions in the world (see Patsiorkovski et al. 1991). In 1971, for example, the mandatory minimum level of education in rural Russia was ten years. In surveys of two rural villages in Russia in 1991 and 1993, the average level of education for adults was almost nine years (O'Brien et al. 1998b: 83).

Persons who have registered as *fermery* have a higher average level of education than *krest'ianskie khoziaistva*, which is largely due to the fact that the former are more likely than the latter to come from urban areas (Wegren 1998:10). The higher level of education of the *fermery*, however, is offset by the fact that they usually do not have much personal experience in agriculture. The high rate of failure of enterprises registered as *fermery* shows that education, by itself, may not be a very significant factor in differentiating between more and less successful small-scale farming operations *at the present time*.

The ability of households, which have remained as *krest'ianskie khoziaistva* to increase their output and sales, is due less to education than to their ability to possess *hand labor*. This hand labor is used to care for animals, process meat and dairy products, and transport agricultural products to market. Thus, it is not human capital, in the Western economic sense of that term that is most likely to differentiate between Russian rural households in agricultural production and sales. Rather, the model for understanding the relationship between household labor and economic success is found in studies of peasant societies in which household labor is a function to a significant degree of the social arrangements in the household. In other words, *household labor is, in part at least, a product of the social capital of the household.*

Peasant household labor can be viewed as a form of human capital, in the sense that it can be measured as an attribute of individuals. Thus, for example, the amount of labor in a peasant household can be quantitatively described as a sum of the weighted contributions of individual household members, of various ages to overall household production and sales (see Chapter 3). To assume, however, that the quantity of labor *maintained* by a peasant household can be understood strictly in terms of individualistic motives and action would be very misleading.

Although the peasant household's success in agricultural productivity depends on its ability to access labor, it is crucial to understand that the household itself is not merely an aggregation of discrete individuals who contribute their labor in return for a specific remuneration. Thus, the labor contributed to household production cannot be compared, for example, to a worker in a factory or a clerk in a store. Rather, the contribution of individual members to overall household production, processing and marketing activities is maintained by commitments to higher level values of *trust* and moral authority that are collective rather than individualistic in nature.

In this respect, the social organization of labor in the peasant household is like the social organization of labor in the highly successful ethnicity-based small businesses in America, which are termed "ethnic enterprises" (Fugita and O'Brien 1991:47-62; Light 1972; Wilson and Portes 1980; Sanders and Nee 1998). In this type of relationship, non-economic bonds of attachment and obligation provide the "moral glue" that keeps individuals contributing long hours of work at what would be considered "exploitive" wages in an individualistic worker-employer contractual relationship. Moreover, in this type of situation, the remuneration the individual household member receives is totally dependent on the collective success of the production unit, the household (Fugita and O'Brien 1991:47-62; Light 1972). Netting describes the cultural supports for peasant household social and economic organization:

> Smallholder agriculture helps to restore or maintain a common focus on the domestic functions, concentrating the time and effort, the rights and duties of household members so that the density and significance of shared, cooperative activities are increased. *The household does farm work, but not as separate*

and independent producers...Neither are household members equivalent to hired laborers, serfs on a manor, slaves on a plantation, specialized employees of an agribusiness, or members of a collective, and in fact these agricultural institutions seem poorly adapted to a regime of intensive cultivation... (1993:59) (our emphasis).

Netting goes on to elaborate the way in which smallholder enterprises, like the Russian peasant household, are rationally organized and efficient economic organizations and not merely cultural relics from the past. Because the social organization of labor is based upon mutual trust and dependence, rather than individualistic wage-labor contracts, these types of enterprises are able to substantially reduce "transaction costs" involving relationships between workers and "monitoring" of work performance (1993:71-74). The contributions of individual household members to the collective household "public good," which is production and sales, are based on informal norms and sanctions, which as Olson has shown can be very effective ways of maintaining individual contributions to non-divisible collective goods in a tight-knit small group (1971:53-65).

Thus, it is important for the reader to understand that the social organization and culture of the household generate labor in the peasant household. The specific type of human capital that is most crucial to the economic performance of the peasant household, hand labor, is a function of social capital in the same sense in which the specific type of human capital most relevant to industrial societies, education, is partially a function of the social capital of the household (Coleman 1988).

A classical approach to understanding the social bases of labor in peasant households is found in Chaianov's (1966) work on peasant households in Russia during the early part of the twentieth century. He showed that the availability of labor in the household was strongly connected to a household's capacity to work the land and to produce agricultural products. More recently, scholars have shown that family size, along with class variables, is associated with the differentiation of peasant households in contemporary Third World countries (Deere and de Janvry 1981).

This is not in any way meant to suggest that peasant household production is an alternative to modern agricultural techniques or that small scale peasant household production will remain a significant part of Russian agriculture in the future. Rather, given the current impasse over land reform and the persistence of the large enterprises in capital intensive production, it is likely in the near future, at least, that some portion of peasant households will continue to occupy a niche in which they provide products to local markets. Because the production, processing, and marketing of these products are very labor intensive, we would expect that the relative success of a peasant household in this type of enterprise will be influenced by its ability to obtain household labor.

There are, however, two important ways in which human capital formation in Russian peasant households today is different than it was in Chaianov's time. The first is that the social organization of the contemporary peasant household and its production and sales of agricultural products is very much connected to the economic situation of the large enterprises, the *kolkhozy* or *TOOs* in which working age adult household members work. The second is that contemporary peasant households can access labor from urban centers.

Earlier, in Figure 1.1 in the Introduction, we presented the institutional structure of rural areas during the Soviet era. With the exception of the period of the New Economic Policy (NEP) in the 1920s, the overriding goal of Soviet agricultural planners was to create a system modeled after large-scale urban industrial factories. This, combined with the Bolshevik's ideological aversion to private property, provided the rationale for the collectivization of agricultural land and the persecution of more successful peasants who were called "*kulaks*" (Fitzpatrick 1994). The cornerstone of the Soviet model was the *kolkhoz* and the *sovkhoz*. *Krest'ianskie khoziaistva* provided labor for the *kolkhozy* and *sovkhozy* (Van Atta 1993). Moreover, the *kolkhozy* and *sovkhozy* became the dominant service providers in rural areas. They provided not only the means for agricultural production but also full or partial support for a whole range of services, including shops for the purchase of food and household supplies, medical services, household utilities, schools, and maintenance of roads (O'Brien et al. 1993; Patsiorkovski 1991).

The Soviet government permitted each peasant household to retain a small private plot. *Kolkhozy* and *sovkhozy* administrators assisted peasant households with their plots, including help with cultivation, fertilizer and sales of products, but, production on these private plots relied primarily on the hand labor of household members, kin from other places, and neighbors (O'Brien et al. 1993).

Although Soviet leaders oftentimes were frustrated by what they saw as competition between *kolkhozniky* (collective farmers') work on their private plots and their work on the *kolkhozy*, they tolerated these small parcels of private land because they remained an essential part of overall

28 *Household Capital and the Agrarian Problem in Russia*

agricultural production. At the same time, however, the government did not provide peasant households with the wherewithal to improve them, lest they compete with the *kolkhozy* or *sovkhozy*. The inability to gain access to larger tracts of land, the lack of a fully developed market system, as well as, restrictions on ownership of animals and machinery, limited the extent to which peasant household production could expand. Thus, the Soviet system did not encourage the development of additional human capital in the household (Fitzpatrick 1994; Humphrey 1988:61-63; Van Atta 1993b).

The introduction of market principles into the Russian economy in the late 1980s, however, created new incentives to increase human capital in rural households. Government subsidies to *kolkhozy* and *sovkhozy* declined while, at the same time, the central government retained monopolies over food processing facilities, which meant that they received less cash for the raw agricultural products they produced. This, in turn, meant that the *kolkhozy, sovkhozy,* and *TOOs* were not able to maintain income payments to their workers (*kolkhozniky*). At the same time, the Russian economy experienced very high rates of inflation. In short, rural households were under extreme pressure to find new ways to generate additional income.

Even before the collapse of the Soviet Union, the proportion of household income derived from the *kolkhoz* was declining, from 58.6 percent in 1990 to 45.3 percent in 1991. At the same time, the percentage of total household income derived from private household plot production was increasing, from 21.5 percent in 1990 to 30.0 percent in 1991, and 42.0 percent in 1992 (*Narodnoe Khoziaistvo Rossiiskoi* 1993: 147, 162).

The expansion of peasant household production follows more closely an "informal economy" model rather than that of the formal legal designation *fermer*. Russian government taxes on officially registered private farmers are extremely high, as they are on all private businesses, and thus there are disincentives for a household to officially register for this new category. In addition, official registration as a *fermer* means that the household cannot get certain kinds of assistance from the *kolkhoz* or *TOO* which their neighbors receive. In interviews in Yekaterinburg the researchers found that many individuals who became officially registered as private farmers in 1992 or 1993, officially removed themselves from this status in 1994 or 1995. They were continuing to build their household production, processing, and marketing capabilities, adding livestock, equipment, and buildings, and, in some cases food processing, but they were now officially returning to the traditional status of *krest'ianskoe khoziaistvo*.

Moreover, in order to provide incentives for peasant households to continue to provide labor, in the absence of wage payments, *kolkhozy* and *TOO*s have relied on payments in the form of grain, which is used by peasant households to feed their increasing stocks of animals. In addition, *kolkhozy* and *TOO*s often provide households with low cost building supplies to increase the size of their homes and structures for keeping animals or producing value-added products, as well as, very expensive material infrastructure which households would not be able to provide on their own. In one village (Vengerovka) in this study, for example, the *TOO* provided the materials and machinery to build a natural gas line which was connected to the state pipeline. This provided an inexpensive source of heat for peasant households and enabled them to keep larger numbers of animals during the winter months.

In short, conditions in the post-Soviet period have created incentives to increase household production. Because of a lack of mechanical equipment, most peasant household production depends, in whole or in part, on hand labor. Thus, the addition of working age adults greatly expands a household's productive capacity.

In Chaianov's time, the primary way in which a peasant household could increase its human capital was by adding new members through birth. Thus, his formula for estimating the economic strength of the peasant household placed primary weight on the life cycle of the household. This association remains valid to a certain extent to this day, as in the case where younger couples with children who can help out on the plot are clearly at an advantage over aged retired couples or widows. But, there are other ways in which the supply of labor in the household can be increased today. This is seen in a comparison of household types in Russian villages in 1991, 1993, and 1995. Extended family households, other than adult children and their parent(s), for example, made up only 4.8 percent of the households in a survey of Russian villages in 1991. By 1993, however, this type made up 12.7 percent of the households in a similar Russian village survey and by 1995 this type of household accounted for 17.3 percent of all households in the villages surveyed (O'Brien et al. 1996a; Patsiorkovski and O'Brien 1996:145).

Moreover, the size and age composition of this type of extended family household has changed during the past few years of economic and social change in the Russian villages. In 1991, these households typically were made up of inter-generational families, such as an aged mother in her eighties and her son in his fifties, or a brother, sister or niece living with a middle-aged couple who did not have any children living at home. The average size of this type of household in 1991 was 2.5 persons. In 1995, 62.5 percent of the households of this type had three or more family members, 33.9 percent of them had four or more family members, and 10.1 percent had five or more members. Almost all of the increase in the size of this type of household has been by the addition of existing members from the village or relatives from the city (oftentimes displaced defense industry and military workers) and not from new births within the village (*Uroven' zhizni naselenia Rossiiskoi Federatsii* 1995:10).

Demographic trends show that although the mean age of the rural population in Russia remains high, it is similar to what is found in rural regions of the American Midwest (Patsiorkovski and O'Brien 1996:136).

There has been a substantial growth of population in rural Russia since 1991, as illustrated in Graph 2.2. The peak of this increase occurred in 1995, and even though the rural population has dropped slightly in 1996 and 1997, it is still considerably higher than in 1991 (see Chapter 2). Wegren (1998) attributes this increase to two sources. He says:

> . . . Moreover, the first was an urban-rural movement, no doubt motivated by the decline in urban living standards. The second source stemmed from a more general inflow of persons from the 'near abroad' in the wake of the collapse of the USSR. Due to the high cost of urban housing, as well as competition for jobs and increasing unemployment, arrivals from the near abroad are not settling in Russian cities, but rather in the countryside. We should note, however, that an estimated 80 percent of new arrivals from former Soviet republics are former urban residents, and thus it is entirely possible that they will not settle in the countryside permanently.
>
> Overall, significant percentages of new arrivals into rural areas are less than 40 years of age. For example, during 1995, rural areas in Russia experienced a net inflow of 29,289 persons aged 20-29, and another 43,179 aged 30-39. These two age groups accounted for 54 percent of net migration into rural areas in 1995. Overall, persons of working age accounted for 44 percent, and persons too old to work accounted for 21 percent of net rural migration (1998:23-24).

Graph 2.2: The Size of the Rural Population in Russia from 1989 to 1997 (in millions)*

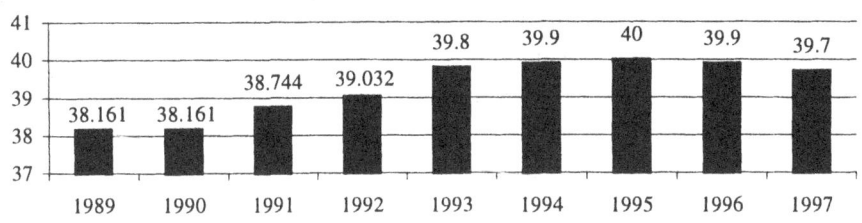

*Source: Goskomstat (1990:16); Goskomstat (1992:22); Goskomstat (1997:17).

Social Networks, Community Attachment, and Peasant Household Production and Sales

Household labor, as described above, is a form of human capital that is produced and maintained by social capital. Two "purer" forms of social capital that are likely to affect peasant household production are social exchange helping networks and community attachment. These are referred to as purer forms of social capital since their measurement is made by observing, in Coleman's words, "some aspects of social structure" (Coleman 1988:S98) or "social infrastructure" (Flora and Flora 1993:49) which create competitive advantages for individuals, organizations, or communities. These purer forms of social capital are rooted in specific types of social relationships that provide specific advantages with respect to accessing information, credit, or labor. Some ethnic communities, for example, have an advantage in developing small businesses because they have "quasi-kin" social networks that support rotating credit associations. These associations provide startup funds for such businesses. In addition, some of these ethnic communities have created social networks that support vertically integrated production, processing, and sales (Light 1972; Fugita and O'Brien 1991:47-62; Wilson and Portes 1980). Flora and Flora (1993) have identified symbolic diversity, resource mobilization, and quality of networks, as three types of social capital which are associated with success in community development efforts in rural places in the United States (see also O'Brien et al. 1991; O'Brien et al. 1998c).

Like all authoritarian regimes, the Soviets saw intermediary structures between the individual and the state as potential competitors (Kornhauser 1959) and thus attempted to destroy traditional types of voluntary associations, informal marketing and trading networks, and other forms of community attachments which exist in rural areas in other countries. Nevertheless, the Soviets did not entirely destroy all forms of informal cooperation and networking. Soviet citizens, in general (Slapentokh 1989), and rural villagers, in particular (Dershem 1995), came to develop and maintain informal social support and material helping networks which permitted them to cope with the command economy.

Our second hypothesis can be stated as follows:

H-2: *Peasant households with larger helping networks, will have an advantage over other households in (a) agricultural production and (b) sales.*

It is expected that new opportunities for production and sales will encourage households to develop personal social helping networks to assist them in various phases of these activities. Nevertheless, it is also expected that some households will have more success than others will in developing and maintaining these networks and that this, in turn, will provide them with an advantage in competing in the new market economy.

Our third hypothesis can be stated as follows:

H-3: *Households that are more integrated into village life will have higher levels of (a) productivity and (b) sales.*

The Soviets never entirely destroyed peasant household attachments to their villages (Humphrey 1988). Studies have shown that the extent to which a household is integrated into the village will be associated with its support for privatization and market reform. Households that have sources of income other than the *kolkhoz* or *TOO* are also more likely to be integrated into broader village life. Thus, we might expect that these households will be better able to use this form of social capital to facilitate their own household production (O'Brien et al. 1993; O'Brien et al. 1998b; Patsiorkovski and O'Brien 1996).

Types of Physical Capital Through Which Household Labor, Social Networks, and Community Attachment Affect Household Production and Sales

The first three hypotheses describe expectations about the *total effect* of household labor, social networks and community attachment on the production and sales of peasant households in Russia. Our longer-range goal, however, is to identify specifically *how, in a causal sense,* the different types of capital create advantages for some households. Here the goal is twofold. First, we want to identify how a specific type of household capital works *directly and indirectly* to increase production and sales. Second, we want to identify ways in which specific types of household capital have *multiplicative* effects; that is, advantages which multiply in specific ways as the household goes through the process of producing and selling products from their plots.

The possible causal linkages between household labor, social networks and community attachment and household production and sales are shown in Figure 2.1.

Figure 2.1: Conceptual Model of Possible Direct and Indirect Effects of Household Labor, Social Networks and Community Attachment on Peasant Household Production and Sales

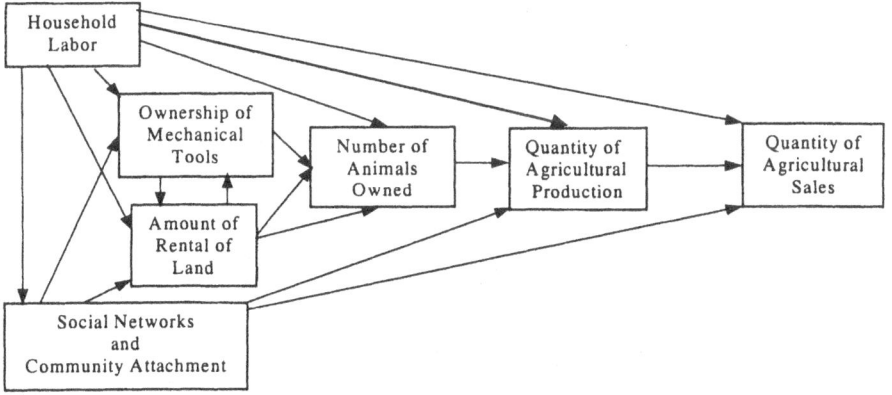

Our fourth hypothesis can be stated as follows

H-4: *Household labor, social networks and community attachment will have direct effects on household (a) production and (b) sales.*

The fourth hypothesis is simply based on the notion that additional working hands (human capital), and social networks, community involvement and community attachment (social capital) will be positively associated with a household's ability to produce and sell more goods. An additional concern, however, is the manner in which specific types of capital may affect household production and sales *indirectly*, thereby increasing the number of mechanical agricultural tools purchased, the amount of land rented from the Village Council and the number of animals owned by households.

Mechanical Equipment

During the Soviet era, the *kolkhozy* and *sovkhozy* maintained almost a complete monopoly over agricultural technology. Peasant households were dependent on the *kolkhoz* for assistance in the cultivation of their plots, access to fertilizers and other inputs, as well as transportation to market (O'Brien et al. 1993). In addition, the Soviet policy of creating very large but highly specialized agricultural units meant that processing of raw materials, especially meat and dairy products, took place in large plants very far removed from the local village and *kolkhoz*. There were almost no opportunities during the Soviet period for peasant households to gain access to their own mechanized agricultural equipment to increase their production potential, and most importantly, no opportunity to produce value-added products (Van Atta 1993).

At the beginning of the post-Soviet period, the opportunities for expansion of peasant household plots were associated with access to relatively simple hand mechanical tools which could turn raw agricultural materials into value added products. This included cultivation implements which could be used with tractors that were borrowed from the *kolkhoz* or *TOO* and equipment for separating cream from milk, an important step in the development of value-added dairy products, equipment for making hams and sausages, and sprayers for fruit trees. Households that have mechanical equipment will have higher levels of productivity. The more human (number of working-age adults) and social capital it possesses, the more likely it is that that household can pay for the costs of that additional capital expenditure. In addition, we would expect that households with more extensive social networks and community attachments would be more likely to obtain information about opportunities to obtain mechanical equipment.

Thus, our fifth hypothesis can be stated as follows:

H-5: *Households with higher levels of household labor, social networks and community attachment will have advantages in overall (a) production and (b) sales through their increased ability to purchase mechanical equipment.*

Renting Land from Village Government and Neighbors

As noted earlier, the impasse over land reform at the federal level has severely limited the transfer of land and the creation of a true land market (see Appendix 1 for a chronology of land reform laws). Nonetheless, formal and informal rental arrangements have been taking place in the rural areas of Russia since the late 1990s. Formal rental arrangements are between households and the local village government, which is now the owner of a portion of arable land. None of the households rented land from the *kolkhoz* or *TOO* (see Chapter 7).

Informal arrangements involve a household entering into an agreement to cultivate crops or raise animals on the private plot of one of its neighbors in exchange for giving the neighbors some portion of the products raised on their land. Typically, the amount of land which is rented in these types of informal agreements, is not very large; usually not more than one-third of a hectare is involved in any single transaction. Nonetheless, the addition of even a small plot of land can make a significant difference in the output of labor intensive agriculture, especially when it involves the production of a value-added product, such as hams, sausages or sour cream. We would expect that households with more extensive household labor, social networks and community attachment will have an advantage in this regard, both in terms of having additional hands which can work rented land but also in terms of social networks and community attachments which can lead to the development of social relationships upon which such arrangements are based.

Thus, our sixth hypothesis can be stated as follows:

H-6: *Households with higher levels of household labor, social networks and community attachment will have advantages in overall (a) production and (b) sales through their increased ability to rent land from village government and neighbors.*

Number of Animals Owned by a Household

The most labor-intensive aspect of peasant household production in Russian villages is the care of animals. The number of animals a household can keep is directly proportional to the labor it can devote to their maintenance. Cattle, hogs, sheep and fowl have to be fed and sheltered on a daily basis. This involves moving them from household sheds and barns to fields and ponds each day and then returning them to the household at night. In the case of cows, this involves milking twice a day. All of these activities are especially difficult during the Russian winter, in which feed often has to be heated. Thus, households with more working-age adults can be expected to have an easier time handling a larger number of animals. In addition, having extensive social networks and community attachments will increase the number of persons outside of the household who can be depended upon to help care for animals.

Thus, our seventh hypothesis can be stated as follows:

H-7: *Households with higher levels of household labor, social networks and community attachment will have advantages in overall (a) production and (b) sales through their ability to care for a larger number of animals.*

Household Capital, Income and Acquisition of Durable Goods

Household production and sales create both monetized and nonmonetized income. As used here, monetized income is found in currency and may be used directly to purchase durable goods. Nonmonetized income is found in agricultural products that are generated by a household but which are either consumed by them or exchanged in barter with other households. In order to get a complete estimate of material well-being, including estimates of poverty and vulnerability, it is necessary to include nonmonetized as well as monetized income in assessing a household's position vis-à-vis other

Peasant Households and the Transition 37

households' income (Dershem and Gzirishvili 1998; Rose and McAllister 1996). An important implication of this line of reasoning is that households which lack the resources to obtain monetized and nonmonetized income are the most likely to become defined as vulnerable and thus dependent on social services provided by government or private charity organizations.

Our eighth hypothesis can be stated as follows:

H-8: *Households with higher levels of household labor, social networks and community attachment will have higher levels of both monetized and nonmonetized income.*

Thus, our ninth hypothesis can be stated as follows:

H-9: *Households with higher levels of household labor, social networks and community attachment will own more durable goods.*

The relationships described in hypotheses 8 and 9 are illustrated in Figure 2.2.

Figure 2.2: Conceptual Model of Possible Causal Effects of Household Labor, Social Networks and Community Attachment on Household Income and Acquisition of Durable Goods

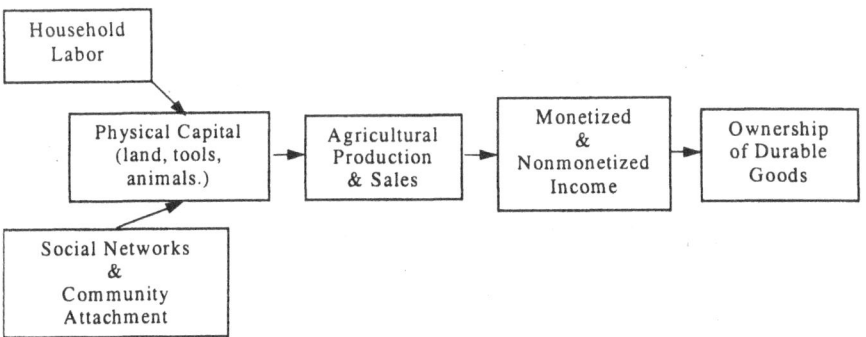

38 Household Capital and the Agrarian Problem in Russia

Modeling the Effects of Household Labor, Social Networks and Community Attachment on Subjective Quality of Life and Stress

Figure 2.3 is a conceptual model that shows the possible direct and indirect effects of human and social capital on stress and subjective quality of life.

Figure 2.3: Conceptual Model of Possible Effects of Household Capital on Subjective Quality of Life and Depression

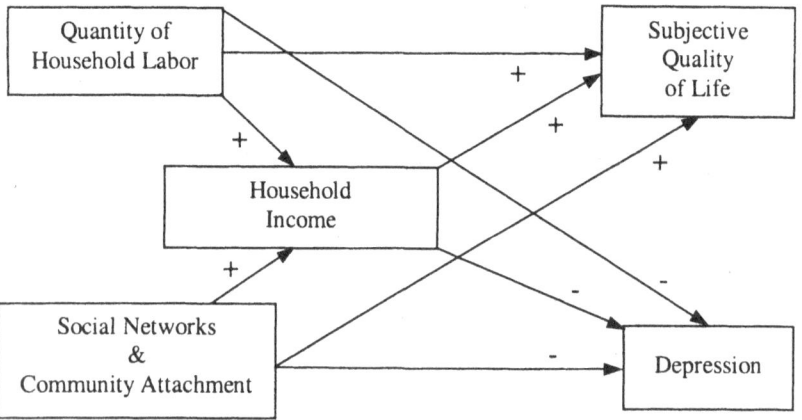

Our tenth hypothesis can be stated as follows:

H-10: *Household labor will have a negative effect on (a) depression and a positive effect on (b) subjective quality of life.*

This hypothesis suggests that having more working-age adults in the household will reduce symptoms of depression and improve subjective quality of life. It is expected that having more household labor will instill a sense of security among household members. Studies in North America have shown that size of household is associated with a depressed mood, since household income is spread over a larger number of individuals in the household. Moreover, household income in the United States typically is operationalized in terms of household size (Ross and Huber 1985). There are, however, several reasons why we would expect that a larger number of *working age adults* in a rural Russian village household should not have this effect. The psychological distress currently found in Russian villages is associated with the pressures that have been brought on by the worsening conditions of the *kolkhozy* and *TOO*s. This includes the reduced capacity of these organizations and the central government to provide social service support, the increased demand on peasant households to generate income from their household plots, and the overall strain of inflation (which currently is under control but not before prices have risen dramatically). Thus, any conditions that help household production also help in dealing with the psychological distress associated with economic and social change. Households with more working age adult members should experience lower levels of psychological distress in getting basic tasks related to agricultural production completed, as well as, feel more secure about their ability to increase production in order to keep up with the increased demands of a market economy. It is expected that the benefits of having more human capital in the household will outweigh what are likely to be "transaction costs" associated with dealing with a larger number of interpersonal relationships.

Our eleventh hypothesis can be stated as follows:

H-11: *Social networks and community attachment will have (a) negative effects on depression and (b) positive effects on subjective quality of life.*

Social exchange and helping networks have been found in other settings to facilitate coping with difficult situations, especially when these situations put increased demands on the household for time, energy and material resources (Lin et al. 1986). Other studies have found that community attachment is positively related to mental health for rural residents in the American Midwest (O'Brien et al. 1994). Thus, we would expect that these forms of social capital would have a direct positive effect on mental health in the household. In addition, the positive association between social capital and household production should produce an indirect positive effect on psychological mood in the household. As in the case of household labor, it is expected that there will be "transaction costs" associated with maintaining social networks and community attachments but that these costs will be offset by the positive benefits derived from these forms of social capital.

Village Differences in Support of Household Capital

Differences between villages in support of the development of peasant household capital can stem from differences in the policies of their respective *oblasts*, but they can also be the result of differences in cultural and social history and contemporary village leadership. In rural Russia, village leadership is heavily dependent on the past social organization of families and kin, but it also depends upon past and current relationships between the large enterprises, the *kolkhozy* and *TOO*s, and peasant households (Dershem 1995; Dershem 1998). In some villages these relationships were very positive during the Soviet period, whereas in other villages they were not. Even within a single village, the relationships between households and large enterprises can change dramatically when a manager is replaced. In the villages involved in the joint Russian-American research project, for example, large enterprise managers have differed markedly in their level of interest in and support for peasant households. This ranges from situations where the manager does not live in the village and has little interest in social services and everyday problems of households to another situation where the manager is very much connected to other residents, having deep roots and a commitment to household and village viability (Dershem 1995; Dershem 1998).

The social organization and culture of a village can support the development of both types of household capital in two ways. First, it can facilitate access to materials which will assist in the development of the physical structures in the peasant household, the home, the barns, and sheds. Second, it can support the development of informal "networking" between households that, in turn, will facilitate the development of household enterprises.

By the same token, managers of large enterprises can be more or less supportive of the development of household labor, social networks and community attachment in peasant households. This can include assisting households in the development of their physical infrastructure, through the large enterprise's access to wholesale purchases of materials, support for informal networking and exchange relationships between the household and the large enterprises, and by supporting changes which devolve power from the large enterprises to households and local government.

Our twelfth hypothesis can be stated as follows:

H-12: *Village differences in social organization and large enterprise leadership will produce differences in the development of human and social capital in peasant households and, in turn, the overall (a) production and (b) sales of agricultural products by these households.*

The twelve hypotheses just stated will be tested on primary sample survey data collected from household respondents interviewed in a three-wave panel study from 1995 to 1997. Structural equation models will be used to test each of these hypotheses. In addition, pertinent secondary data will be used to identify linkages between our micro-level findings and macro-level household trends in rural Russia.

Summary

The basic analytical unit in this study is the *krest'ianskie khoziaistvo*, the Russian peasant household. Although the average size of land holdings by households of this type is quite small, they account for a considerable portion of the meat, dairy, and vegetable commodities produced in Russia. Moreover, in the post-Soviet period institutional changes have been occurring in Russian villages that are creating opportunities for these households to expand their production and sales.

The primary purpose of this study is to examine the overall economic and psychological adaptation, and subjective quality of life, of rural Russian households to an emerging market economy and to explain the emerging economic differentiation of these households. The theoretical approach to peasant household economic development, as well as the inequalities between rural households focuses on the extent to which individual households and villages have various forms of household capital. The specific types of household capital that are examined include, household labor, which is a form of human capital embedded within household social capital, and two purer forms of social capital, community attachment and social exchange helping networks. It is expected that differences in levels of these forms of household capital will affect the mental health and subjective quality of life, as well as the material well being of rural households. In addition, it is expected that there will be differences between villages in support of the development of household capital that will produce differences in average levels of rural household adjustment to an emerging market economy.

3 Population and Household Structure in Rural Russia

Valeri V. Patsiorkovski, Larry D. Dershem and David J. O'Brien

Introduction

The main concern of this study is to identify how household capital contributes to production and sales, income and access to material goods, and subjective quality of life and mental health in rural Russia. A vital ingredient of household capital is the availability of labor. Examining the structure of rural households is essential to understanding the larger issue of human capital and agrarian reform in rural Russia. As will be recalled from Chapter 1, household labor, which is a form of human capital, is embedded within the structure of the household, which is a form of social capital. This chapter examines changes in the rural population in the eleven major economic or "macro" regions of Russia that have affected the size, structure, and availability of household labor.

Many academic and public policy discussions about Russian agricultural reform have raised the question, is there suitable human capital to support a sustainable Russian agriculture that is part of an international free market agricultural system? There have been three basic responses to this question, each reflecting a different scholarly and public policy point of view. The first perspective views all aspects of Russian rural life, cultural, economic, and social as obstacles to agricultural reform (Nicolsky 1996). The second view is that specific demographic conditions have resulted in the absence of a sufficient number of able-bodied persons in the countryside to support a sustainable agriculture that can compete in a free-market (Wegren 1998:22-23).

A third view is that rural residents in Russia possess deeply ingrained cultural attitudes that are resistant to privatization of land and agricultural reform (Vinogradsli 1997; Martynova 1996). All of these views, suggest that the only way to reform Russian agriculture is to bring into the countryside younger and more educated persons who are not held back by the age and cultural traditions of the peasant households.

44 Household Capital and the Agrarian Problem in Russia

Nevertheless, as will be presented in this chapter, the actual and potential human capital in the Russian countryside is much greater than most scholars have assumed, in terms of age, education, and cultural receptivity to expanding private agriculture. The distribution of this capital varies considerably, however, from one macro-economic region of Russia to another.

Change in the Size of the Rural Population

Overall, the total population of Russia has declined in last five years. In 1992 there were 148.7 million people in Russia dropping to 147.5 million people in 1997 (Russia in Figures 1997:17). Public opinion has expressed shock over this decline, but for professional demographers, it was not unexpected. "Recent demographic trends in post-Soviet Russia have profound roots in the Russian history of the 20th century; in particular they are closely linked to the evolution of the Russian family in the post-war period" (Vishnevsky 1996). It is this historical situation, as well as the current economic crisis that has produced the so-called demographic crisis in Russia today.

Over the last thirty-five years, the rural population in Russia has changed considerably. In 1960, 55.3 million people lived in rural areas, accounting for 46.5 percent of the total Russian population. By 1997, there were 39.7 million people living in rural areas, accounting for 26.9 percent of the total Russian population (see Table 3.1). Since 1992, the size of rural population increased by approximately one million, from 39.0 million to 39.7 million people, as shown in Graph 2.2 (Goskomstat 1997b:Table 2.1). There are, however, large differences in the demographic characteristics of rural areas in different Economic Regions (*Economicheskii raiony*), *oblasts* and *Krays*. These differences reflect the enormous size of the Russian Federation that contains substantial variations in natural ecology and ethnicity (see Appendix 2). Therefore, the distribution of household and human capital in rural Russia varies considerably from one macro-economic region to another.

In 1997, five of the eleven Economic Regions in Russia had a greater proportion of their population living in rural areas than the national average (see Table 3.2). These include the Volga-Vyatka, Central Black-earth, North-Caucasus, West Siberia and East Siberia Regions. The Central

Black-earth and the North-Caucasus Regions had the highest proportion of their population living in rural areas (38.1 and 44.5 percent, respectively). The Central Black-earth Region is the site of the village of Vengerovka, in Belgorod *Oblast* and the North-Caucasus Region is the site of the village of Latonovo in Rostov *Oblast*. These are two of the three villages in the panel study (see Chapter 4).

Five of the eleven Economic Regions had a lower proportion of their population living in rural areas than is found in the country as a whole in 1997. These are the North, Northwest, Central, Ural and Far East Regions. The lowest proportions of the population living in rural areas are in two Economic Regions, the Northwest (13.3 percent) and the Central (17.0 percent). The third village in the our panel study (Bolshoe Sviattsovo, Tver *Oblast*) is in the Central Region and thus provides an important contrast with the two villages in the more viable Central Black-earth and North Caucasus Regions. The proportion of the population in the Volga Region is approximately the same as the proportion of the rural population in Russia.

Change in Rural Population Size by Economic Regions

The changes in the characteristics and sizes of the populations of the different Economic Regions during the past decade, as shown in Table 3.1, indicate that there have been considerable differences in regional demographics and considerable internal migration in Russia. There are four different patterns of rural population change during this time: (1) rapid decline, (2) slight decline, (3) slight increase and (4) rapid increase.

46 *Household Capital and the Agrarian Problem in Russia*

Table 3.1: Components of Rural Population Size Changes (in thousands)

Year	Population at the beginning of year	Total annual increase (%)	Rural proportion of Russian population at end of year
1960	55306.0	-1.2	46.5
1970	49098.1	-1.8	37.8
1979	42177.1	-1.3	30.7
1989	38974.9	-0.5	26.4
1990	38802.3	-0.1	26.2
1991	38744.3	-0.1	26.1
1992	39031.9	1.8	26.3
1993	39753.1	0.4	26.7
1994	39904.0	0.2	26.9
1995	39968.9	0.1	27.0
1996	39855.2	-0.1	26.9
1997	39708.7	-0.1	26.9

Source: *Demograficheskii ezhegodnik Rossii*. Moscow: Goskomstat of Russia, 1997a, pp.21-23.

Rapid decline In five Economic Regions the rural population has declined rapidly in recent years: the North, Northwest, Central, Volga-Vyatka and Far East Regions. Over the last ten years, these Economic Regions have lost more than one-half a million people (548 thousand) from rural areas, for a total loss of 4.3 percent of their rural population (Goskomstat 1997: 30).

Slight decline Only one Economic Region has experienced a slight decline in its population, the Central Black-earth Region. The total size of the population of this region increased but the rural population decreased slightly. From 1989 to 1997, this region lost 2.6 percent of its rural population (Goskomstat 1997:24-32).

Slight increase Three Economic Regions, the Volga, the Ural, and the East Siberia Regions, and the Kaliningrad *Oblast* enclave experienced a slight increase in the size of their total population, including their rural population (2.8 per cent) in the last ten years. From 1989 to 1997, the rural population in the Volga, Ural, and East Siberia Regions increased by 3.7, 2.0, and 1.4 percent, respectively (Goskomstat 1997:31-32).

Population and Household Structure 47

Table 3.2: Proportions of the Rural Population by Economic Regions of Russia from 1970 to 1997 (in percent)

Russia and Economic Regions	1970	1979	1989	1997
Russian Federation	37.7	30.7	26.5	26.9
North Region	35.6	27.5	23.5	24.2
North West Region	20.1	15.4	13.4	13.3
Central Region	28.7	21.4	17.4	17.0
Volga-Vyatka Region	47.2	37.8	31.1	28.8
Central Black-earth Region	59.8	47.9	39.7	38.1
Volga Region	40.6	32.2	26.7	26.9
North-Caucasus Region	50.2	45.1	42.7	44.5
Ural Region	35.4	29.1	25.3	25.6
West Siberian Region	38.6	32.3	27.2	29.0
East Siberian Region	38.3	31.3	28.1	28.6
Far East Region	28.5	25.5	24.2	24.3
Kaliningrad *Oblast*	26.8	23.5	20.9	22.2

Source: *Demograficheskii ezhegodnik Rossii*. Official Publication. Moscow: Goskomstat of Russia, 1997, pp.30- 32.

Rapid increase Two Economic Regions experienced a rapid increase in the size of their total population, including their rural population; the North-Caucasus and West Siberia Regions. Over the last ten years, the rural population in these regions increased by more than one million people on average, or 9.2 percent, whereas the urban population increased by only 2.3 percent in the North-Caucasus Region and by 2.0 percent in the West Siberia Region (Goskomstat 1997:24-32). The rural population in the North-Caucasus Region, for example, decreased up until 1979, but increased by 892,000 from 1979 to 1997. The rural population in West Siberia increased by 7.4 percent in the last ten years. In summary, the large increases in the overall populations of these Economic Regions were largely the result of a rapidly increasing rural population. It is in these regions that household labor has substantially increased.

Change in Population Size by *Oblasts*

The rural population trend in a particular republic or *Oblast* may vary from the general population tendency of the Economic Region in which it is located. Thus, for example, although the Central Black-earth Region has experienced a *slight population decline,* Belgorod *Oblast,* which is located in this region, has experienced a *slight population increase.* Over the last ten years the size of the total population in Belgorod *Oblast* has increased rapidly (10.6 percent) and the size of the rural population size has increased slowly, by 1.1 percent (Goskomstat 1997: 28-31). Alternatively, in the Volga Region, which has experienced an overall increase in total and rural populations, the total and rural population of Penza *Oblast* has declined slightly.

In other *oblasts,* a sharp decline in the rural population has been offset by an increase in their urban population. This is the case, for example, in Briansk *Oblast* in the Central Economic Region where the rural population decreased from 483,000 in 1989 to 462,000 in 1997. The increase in the urban population of this *oblast,* however, resulted in only a slight overall population decline, from 1,475,000 in 1989 to 1,474,000 in 1997 (Goskomstat 1997a:25-31). A similar situation occurred in Kursk *Oblast* in the Central Black-earth Region, in Kurgan *Oblast* in the Ural Region and in Tyva Republic in the East Siberia Region.

In some Economic Regions that have experienced a steady decline in their overall rural population over the last ten years, there have been provinces, which either have had no change or even have had an increase in the size of their rural population. These provinces include Republic of Karelia (in the North Region), Kostroma, Orel, and Yaroslavl *Oblasts* (in the Central Region), Belgorod *Oblast* (in the Central Black-earth Region), and the Republic of Sakha, and Amur *Oblast* (in the Far East Region).

There are different reasons why *oblasts* in various regions have lost rural population. The primary reason for the decline of the rural population in the Northwest, Central and Volga-Vyatka Regions is *their extremely poor material and social infrastructure.* At the same time there is the strong influence of a nearby large city (Moscow, St. Peterburg and Nizhny Novgorod respectively). Higher quality social services and an overall higher quality of life in these urban areas have been a powerful incentive for migration from surrounding rural areas. On the other hand, the decline of the rural population in the Far East Region is associated primarily with a general outmigration of both urban and rural Russians.

The Number and Size of Rural Households

Currently, there are 14 million households in rural Russia. Over the last 30 years, the size of the rural population has decreased by 30 percent (see Table 3.1 and the average size of households has decreased by 21 percent (see Graph 3.1). In spite of these declining figures, the actual number of rural households has decreased by only 5 percent. There are two possible explanations for these seemingly contradictory findings. First, many young rural families have formed new households either by building, buying or receiving a house from a collective farm during this time period. Second, official bureaucratic accounting of households in rural Russia often counts household structures that have no full-time or only part-time inhabitants. Both of these explanations would account for the slight increase in the number of households and the decline in the average size of families. Table 3.3 shows the size of rural households by Economic Regions (see also Appendix 5). Overall, 22 percent of all households in rural Russia contain only one person, while 50 percent of all households consist of three and more persons. The mean difference between the smallest and the largest average size of households in the eleven Economic Regions is equal to almost one-person (0.82).

The average size of households in six of the eleven Economic Regions, North, Northwest, Central, Volga-Vyatka, Central Black-earth and Volga, is lower than the average size of rural households for the country as a whole (2.85 persons). In the other five Economic Regions, North-Caucasus, Ural, West-Siberia, East-Siberia, Far East and Kaliningrad *Oblast* the average size of households are 2.85 persons or larger.

Graph 3.1: Changes in the Size of the Rural Population, Number of Rural Households and Average Size of Rural Households in Russia (1960-1994)

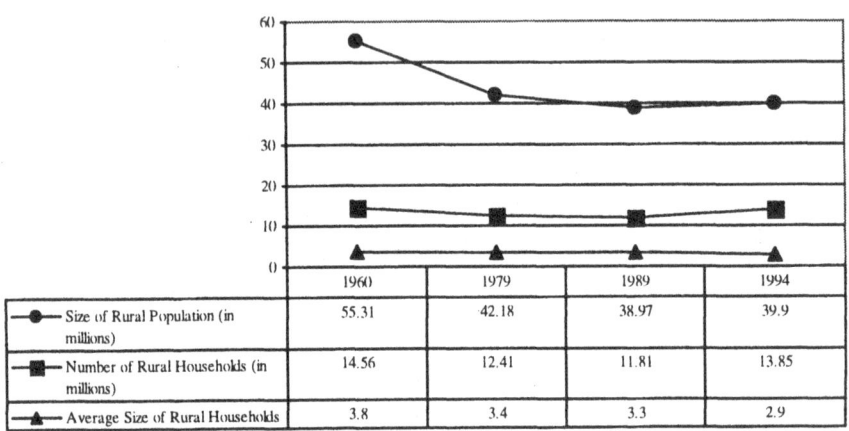

	1960	1979	1989	1994
Size of Rural Population (in millions)	55.31	42.18	38.97	39.9
Number of Rural Households (in millions)	14.56	12.41	11.81	13.85
Average Size of Rural Households	3.8	3.4	3.3	2.9

Sources: *Tipy i sostav domokhosaistv v Rossii po mikroperepisi* 1994 *goda*. Moscow: Goscomstat Rossii, 1995, p. 23. *Demograficheskii ezhegodnik Rossii*. Official Publication. Moscow: Goskomstat of Russia, 1997, p. 23; Anatoly G. Vishnevsky. Family, Fertility, and Demographic Dynamics in Russia: Analysis and Forecast. "Russia's Demographic 'Crisis'". Edited by Julie DaVanzo. Published 1996 by Rand. http://www.rand.org/publications /CF/CF124/CF124.chap1.html.

The smallest rural households, on average, are found in two Economic Regions, the Central (2.45) and the Northwest (2.46). These two areas, which have poor land and have been losing rural population for many years. According to data from the 1989 Russian Census, 805, or 6 percent, of the villages in the Northwest Region and 3,962, or 6 percent, of the villages in the Central Region have been completely abandoned, having no full- or even part-time residents (Goskomstat 1991:68, 360).

The smallest average size rural households are found in Pscov *Oblast* (2.25) in the Northwest Region, and Ryazan (2.25) and Tver *Oblast*s (2.29) in the Central Region. It is this decline in population and average household size, especially, in the Central Region, which is often cited as an example of the decline of labor potential in rural Russia.

Overlooked, however, is the fact that in two Economic Regions, the North-Caucasus and West Siberia, the average size of households is higher than the Russian average (3.27 and 3.18 respectively). Since the North-Caucasus Region contains a large non-Russian Moslem population, such as the Ingushetia (7.13) and Dagestan (4.14) Republics, it is not surprising to find larger than average size of households in this region. However, areas within this region that have predominantly Russian populations, such as Rostov *Oblast*, Krasnodar and Stavropol *Kray* also have, on average, larger rural households; 2.94, 3.12, and 2.99 respectively.

Table 3.3: Number and Size of Households in Rural Russia by Economic Regions (in 1994)

Economic Regions	Number of Households (in thousands)	Per 1000 Households					Average size of Households
		From 1 person	From 2 persons	From 3 persons	From 4 persons	From 5 & more persons	
1	2	3	4	5	6	7	8
North Region	541.0	241	265	192	197	105	2.71
North West Region	439.4	302	286	175	158	79	2.46
Central Region	2111.0	303	294	170	153	80	2.45
Volga-Vyatka Region	945.4	255	266	172	187	120	2.71
Black-earth Region	1217.3	296	301	162	149	92	2.49
Volga Region	1612.1	215	274	185	199	127	2.82
North-Caucasus Region	2251.1	171	233	175	199	222	3.27
Ural Region	1753.7	187	262	183	209	159	2.98
West Siberian Region	1481.8	167	277	199	214	143	2.96
East Siberian Region	828.7	154	247	191	209	199	3.18
Far East Region	601.9	152	232	220	223	173	3.15
Kaliningrad *Oblast*	67.2	187	269	195	200	149	2.96
The Russian Federation	14001.4	220	268	181	190	141	2.85

Sources: *Rossiiski Statisticheski Ezhegodnik* 1996. Moscow:Goskomstat, pp. 718-720; *Tipy isostav domokhosaistv v Rossii po mikroperepisi* 1994. Moscow: Goskomstat, 1995, pp. 23-32.

The opportunity to cultivate high quality land within the latter regions has encouraged the growth of the rural population, thereby increasing household labor potential. At the same time, some *oblast* administrations within these regions have developed new initiatives to support agricultural and rural development. A similar situation is found in other *oblasts* that have predominantly Russian populations. For example, the average size of rural households in the Astrakhan *Oblast* (Volga Region) and Irkutsk

Oblast (East Siberia Region) is 3.18 persons, and in Khabarovsk *Oblast* (Far East) the average size is 3.16 persons.

These marked variations in household size show that it is extremely difficult to generalize about rural Russia, especially with respect to actual or potential household labor. This calls to mind the observation by Russian geographers that Russia is at minimum ten different countries (Martynov et al. 1997).

The Three Villages in the Study

The three villages in the panel study are found in regions that vary considerably in their support of rural populations. Bolshoe Sviattsovo, located in Tver *Oblast* in the Central Region, represents an area with poor soil and a declining rural population (see Graph 3.2). On average, there are 16.2 households, or 37 persons, per village in Tver *Oblast*. Approximately one-half of these households consist of a single retired person or a couple without children. More than 10 percent of these villages had no permanent residents in 1989 (Goskomstat 1991:361). The low level of human capital found in Bolshoe Sviattsovo is typical of what would be found in other rural villages in Kaluga, Ryasan, Smolensk, Tula, Vladimir, and Yaroslavl *Oblast*s in the Central Region and Pscov and Novgorod *Oblast*s in the Northwest Region (Goskomstat 1991:174).

Vengerovka, in Belgorod *Oblast* in the Central Black-earth Region represents a rural area with much better soil and a much more viable rural population. Belgorod *Oblast* contained 1,590 villages and 198,200 households in 1989 (Goskomstat 1991:194). On average, there were 124.7 households, or 320.4 persons, per village. In 1989, only 49 villages in this *oblast* had no permanent residents (Goskomstat 1991:361).

Latonovo, in Rostov *Oblast* in the North-Caucasus Region, represents a rural area with excellent soil and a very large rural population. In 1989, there were 2,274 villages and 470,600 rural households in Rostov *Oblast* (Goskomstat 1991:232). On average, there were 207 households, or 618.9 persons, per village in this *oblast*. Only 10 villages in this *oblast* had no permanent residents in 1989 (Goskomstat 1991:361).

Graph 3.2: The Average Number of Households and Residents in Three *Oblasts* Where the Study Villages Are Located

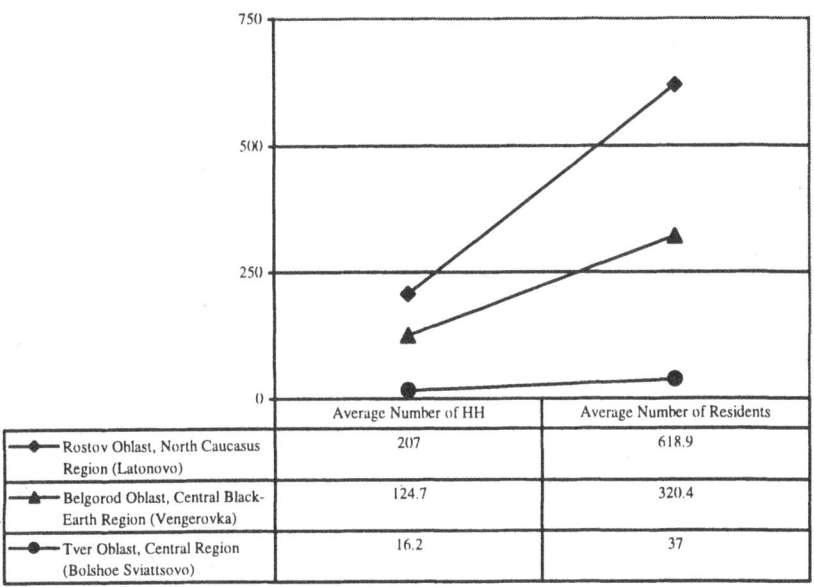

	Average Number of HH	Average Number of Residents
Rostov Oblast, North Caucasus Region (Latonovo)	207	618.9
Belgorod Oblast, Central Black-Earth Region (Vengerovka)	124.7	320.4
Tver Oblast, Central Region (Bolshoe Sviattsovo)	16.2	37

These data show that there are large differences in the average number of households and number of people per village in these three *Oblasts*. Villages in Belgorod *Oblast* have, on average, eight times more households than villages in Tver *Oblast*. Villages in Rostov Oblast have thirteen times more households than villages in Tver *Oblast*.

When considering the number of residents per village, Belgorod *Oblast* contains, on average, nine times more residents per village and Rostov *Oblast* contains, on average, seventeen times more residents per village than Tver *Oblast*. Thus, there are large differences in the potential capacity for the development of household-based agriculture in the three villages in our panel study.

Age Structures

The age structure by gender of the rural population in Russia in 1997 is shown in Table 3.4. This table shows that there is a greater percentage of males than females less than 17 years of age in the age group 64 years and older, however, the proportional representation of men is less than women.

54 Household Capital and the Agrarian Problem in Russia

Approximately 60 percent of men and women are of working age (18 to 64 years of age). This indicates that a large proportion of the current population in rural Russia is capable of working in different types of agricultural enterprises. The age structure in rural Russia today does not suggest a weak demographic base for agricultural production or entrepreneurial activities.

Table 3.4: Age Structure of the Rural Population in Russia in 1997 (in percent)

Age Groups	Male (N=18,909,333)	Female (N=20,880,484)
<5	6.2	5.4
5 – 17	23.8	20.7
18 – 24	10.0	8.1
25 – 44	30.8	25.9
45 – 64	19.9	20.5
65 – 74	7.3	12.3
> 74	2.0	7.1
Total	100.0	100.0

Source: *Chislennost' naselenia Rossiiskoi Federatsii po polu i vozrastu na* 1 January 1997. Moscow: Goskomstat Rossii.1997. p. 246.

Hard physical labor, along with primitive technology and lack of mechanization make the Russian peasant way of life extremely arduous. Although there is some degree of mutual assistance within the household, women are responsible for the majority of tasks related to the upkeep of the household.

Rural households that are comprised of a single man or woman cannot reach the level of agricultural production achieved by other demographic types of families. Furthermore, each additional member of a rural family must contribute to household labor tasks. For this reason, household age structure is a critical element in determining the actual and potential level of household agricultural production.

Table 3.5: The Structure of the Rural Population in Russia by Human Capital Age Groups and Economic Regions in 1997 (in percent)

Economic Regions	Rural Population by Age groups									
	7 & less	8-11	12-14	15-16	17-65	66-70	71-74	75-79	80 & more
North Region	3.19	3.70	3.75	3.65	3.65	3.36	3.50	3.30	2.70
North West Region	1.88	2.33	2.34	2.36	2.73	3.15	3.59	3.44	3.15
Central Region	9.52	10.54	10.46	10.49	12.58	16.14	18.54	17.20	16.73
Volga-Vyatka	5.47	5.85	5.72	5.71	6.16	7.27	8.21	8.67	8.33
Black-earth Region	7.25	6.02	5.84	5.85	7.33	10.60	11.00	10.91	12.84
Volga Region	10.98	10.88	10.60	10.62	11.32	11.96	11.51	11.94	13.46
North-Caucasus	22.91	20.09	19.89	20.12	19.42	16.82	15.32	15.83	16.73
Ural Region	14.00	14.34	14.16	13.94	13.02	12.81	11.41	11.69	11.49
West Siberian	11.12	12.13	12.94	12.95	11.59	9.83	9.29	9.32	8.20
East Siberian	8.17	8.44	8.51	8.52	6.92	5.19	4.87	4.88	4.08
Far East Region	5.02	5.16	5.28	5.28	4.73	2.44	2.34	2.46	2.00
Kaliningrad *Oblast*	0.49	0.52	0.51	0.51	0.55	0.43	0.42	0.36	0.29
The Russian Federation	**100.0**	**100.0**	**100.0**	**100.0**	**100.0**	**100.0**	**100.0**	**100.0**	**100.0**

Source: *Chislennoct' naselenia Rossiiskoi Federatsii po polu i vozrastu na* 1 January 1997. Moscow:Goskomstat Rossii, 1997.

It will be recalled that the average size of rural households in Russia is 2.85 persons (see Table 3.3). In the Economic Regions with fertile soil and potentially high agricultural production (the North-Caucasus, Ural and West Siberian regions), however, 35 to 40 percent of the rural households consist of 4 and more persons (Appendix 5).

There are large differences between Economic Regions in the age structure of the rural population (Table 3.5 and Appendix 6). Graph 3.3 shows that there are five basic types of age structures in rural Russia. Perhaps the most interesting age structure is found in the North Region. This region has approximately 1.4 million people, which is 3.7 percent of the total rural population in Russia (Goskomstat 1997:30). The proportional representation of all age groups is virtually equal. This structure is unusual for Russia because it indicates a contemporary model of simple population replacement. There are two possible explanations for this structure.

Graph 3.3: Five Types of Age Structures in Rural Russia

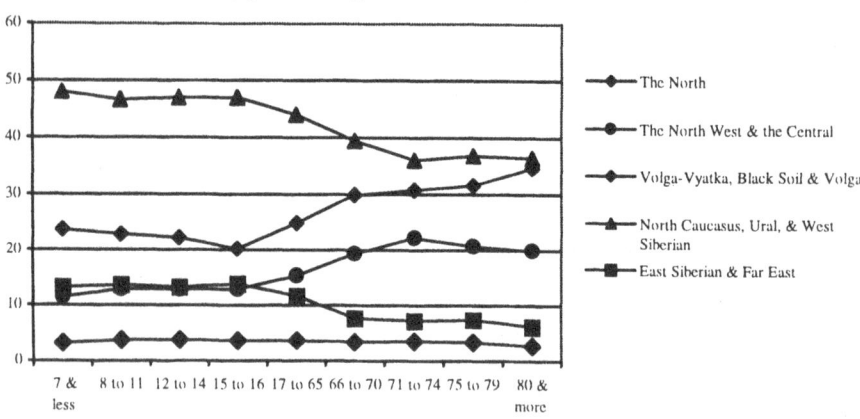

One explanation might be that there is a higher death rate for the elderly population, 66 years of age and older, in this region than in other regions. This, however, is not the case. Russian Census data show that the rate of death for the elderly population in this area is close to the national average. In 1996, there were 16.7 deaths per 1,000 people in the North Region compared to 16.2 deaths per 1,000 people in Russia as a whole (Goskomstat 1997:79). The other, more plausible, explanation is that elderly people have migrated from rural to urban areas. This explanation is supported by statistical data. Over the last 10 years, the net migration rate for the rural population in the North Region is negative whereas the net migration for rural areas in Russia as a whole is positive. In 1996, for example, the average annual net migration per 10,000 of rural population was -68 in the North Region and +9 in Russia as a whole (Goskomstat 1997:541).

The second type of age structure in the rural population is found in the Northwest and Central Regions. These two regions contain slightly more than 6 million rural residents, or 15.4 percent of the entire rural population in Russia (Goskomstat 1997:30). This type of age structure contains a low proportion of children and a high proportion of elderly. The birth rates in the rural areas of these two regions are the lowest in rural Russia. In 1996, the birth rate in these regions was equal to 7.25 births per 1000 people while the average birth rate in Russia was 10.4 births per 1000 people (Goskomstat 1997:79). It is this type of age structure that many specialists refer to when they speak of a "demographic crisis" in rural Russia (Wegren 1998; Smirnov 1996).

Nonetheless, there is a large number of adolescents in these two regions, especially in the Central Region. In addition, during the past few years the net inmigration rates in these regions have been higher than the average annual net inmigration rate in rural Russia as a whole. In 1996, the net inmigration rate in the Northwest Region was 112 persons per 10,000 population, and 44 persons per 10,000 population in the Central Region (Goskomstat 1997:541). This influx of new people into rural areas in these regions is an important compensation for low birth rate. In short, there is some evidence to suggest that a substantial portion of the rural households in these regions have adequate levels of household labor.

The third type of age structure is found in the Volga-Vyatka, Central Black-earth and Volga Regions. These regions contain slightly more than 10 million people, or 25.3 percent of the total rural population in Russia (Goskomstat 1997:30-31). Similar to the second type of age structure type, regions with this type of age structure have a low proportion of children and a high portion of elderly. But this age structure differs from the second type in that its birth rate is slightly higher but the proportion of its rural population that is elderly also is greater than in regions with the second type of age structure.

In 1996, the birth rate was 9.1 births per 1000 population, which is close to the average birth rate of 10.4 per thousand in rural Russia as a whole (Goskomstat 1997:81-83). The higher proportion of elderly in these regions is due to a longer life expectancy than is found in the Northwest and Central Regions. In 1996, the life expectancy at birth was 65.8 years in the Volga-Vyatka, Cental Black-earth, and Volga Regions, whereas in the Northwest and Central Regions it was 63.6 years, and 64.7 years, respectively (Goskomstat 1997:102-106).

There is a relatively large proportion of adolescents in regions of this third type, especially in the Volga Region. The net inmigration rate here is slightly higher than the average annual net migration rate in the whole of rural Russia during last 5 years, but a little lower than the inmigration rate in the Northwest and the Central Regions. In 1996, the net inmigration rates were 11 persons per 10,000 in the Volga-Vyatka, Region, 49 persons per 10,000 in the Black-earth Region, and 14 persons per 10,000 in the Volga Region (Goskomstat 1997:541-542). As in the case of the second type of age structure, the process of migration partly compensates for the low birth rate in these regions.

The fourth type of age structure is found in rural areas of the North-Caucasus Region, the Ural Region, and the West Siberian Region. These regions contain a rural population of 17.5 million, or 44.1 percent of the total rural population in Russia (Goskomstat 1997:31-32). In contrast to the previous three types of age structures, this type contains a high proportion of children and a relatively low proportion of elderly. In 1996, the birth rate in these regions was 12.2 births per 1,000 population (Goskomstat 1997:84-86), which is significantly higher than the average birth rate in rural Russia and in all of the three types of age structures described above.

The low proportion of elderly in regions with this type of age structure is associated with a longer life expectancy, 65.4 years in 1996, which is higher than the life expectancy found in the Northwest and Central Regions (Goskomstat 1997:107-109). At the same time, there is a large proportion of children and adolescents in rural areas of the North Caucasus, Ural and West Siberia Regions. The regions with this age structure have the highest level of rural household labor of any regions in Russia.

These regions with the highest levels of rural household labor contain many non-Slavic ethnic groups, especially Moslem populations that traditionally have large families. Yet, ethnicity does not totally account for the age structure in these regions. Within these Economic Regions there are *Oblasts* in which the majority of the rural population is of Russian descent, such as Krasnodar *Kray*, Rostov *Oblast,* and Stavropol *Kray*. These three locations contain 63.1 percent of rural population in these regions (Goskomstat 1997:31).

The net migration rates in the North Caucasus, Ural and West Siberia Regions is a little higher than the average annual net migration rate in Russia during last 3 years, and a little lower than in the North West and Central Regions. In 1996, the net migration rate in these areas were 6 persons in the North-Caucasus, 29 persons in the Ural, and 13 persons in West Siberia per 10000 population (Goskomstat 1997:542-544).

The fifth type of rural age structure type is found in East Siberia and Far East Siberia. These two regions contain 4.4 million rural inhabitants, or 11.1 percent of the total rural population in Russia (Goskomstat 1997:32). Compared to the previous four rural age structure types, this type has a relatively high proportion of children and an extremely low proportion of elderly. In 1996, the birth rate in these two regions was 12.2 births per 1000 population (Goskomstat 1997:87-88). This is similar to the birth rate for the previous age structure types, and significantly higher than the average birth rate in Russia.

The low proportion of elderly in these regions is due to the low life expectancy in these regions, which is the lowest of all regions in Russia. In 1996, the life expectancy in East Siberia and Far East Siberia was 62.5 years (Goskomstat 1997:109-110). In addition, there as been a higher negative net migration rate in these two regions than in Russia as a whole during the last three years. In 1996, the net migration rate was -58 persons per 10,000 population in East Siberia, and -150 persons per 10,000 population in the Far East Region (Goskomstat 1997:544-545). Nevertheless, there are a relatively large numbers of children and adolescents in these regions, which indicates that the age structure will accumulate higher levels of human capital and household labor as time goes on.

Summary

Reversing a long-term trend, the rural population in Russia increased from 1989 to 1997. This reflects a de-urbanization process that occurred during the transition from the Soviet to post-Soviet periods. During this time, many families preferred to have access to land to grow food for their own consumption. However, some Economic Regions experienced a decrease (North, Northwest, Central, Volga-Vyatka and Far East) while others experienced an increase (North Caucasus, Volga, Ural, West Siberia and East Siberia) during this time period. The rural population increased in those regions that produce the largest portion of agricultural output in Russia. *Contrary to the conventional wisdom about a demographic crisis, there is sufficient household labor to produce growth in agricultural production in enough regions of Russia to develop a viable agricultural economy.*

Not only did the number of individuals living in rural areas increase, on average, in rural Russia but the number of households also increased by 2 million during the first half of the transition period (1989-1994). The average size of rural households decreased, during this same period of time, from 3.3 to 2.9 people. This resulted from a division of extended families into new households. This creation of new households provided these extended families with the ability to access more land, especially communal and rented land from village governments.

There is considerable variation in the amount of rural household labor present in different Economic Regions. Minimums of 45 to 50 percent of the 14 million rural households in Russia have a high level of household labor. Since there is low level of mechanization in peasant household production in rural Russia, hand labor by households still accounts for almost one-half of the total Russian agricultural output. This speaks to the fact that if rural households in Russia are able to obtain land, credit and equipment they will be able to markedly increase their production.

4 Research Design

Larry D. Dershem, David J. O'Brien and Valeri V. Patsiorkovski

Introduction

The most appropriate research design to examine the effects of household labor and social capital on household viability is a multi-wave panel study in which the causal relationships between the dependent and independent variables can be measured over time. The basic panel design in this study is the measurement, *over a three-year period*, of production and sales, income and access to material goods, and subjective quality of life and depressive symptoms in households with different levels of human and social capital. In this design, the same households, and same informants within the households, were interviewed in three waves of interviews, each interview being separated by a time span of one year.

Preparation for the panel study was begun in 1991, at which time, the first of two studies of Russian peasant households was conducted by the Department of Rural Sociology, University of Missouri and the Institute for Socio-Economic Studies of Population (ISESP), Russian Academy of Sciences, in Moscow. A second study was conducted in 1993. The experience of conducting the 1991 and 1993 studies provided the researchers with an opportunity to become familiar with village life, leadership, customs, and social organization, develop training and instrumentation, and gain the trust and cooperation of persons who would become panel respondents. In addition, these surveys provide very important longitudinal indicators which, when combined with the three waves of panel data, show changes from 1991 to 1997. This longitudinal perspective on Russian rural households will be presented along with the panel study findings in the following chapters.

The 1991 survey, which was conducted in two villages, focused on rural residents' level of satisfaction and preferences with social services and various aspects of rural life. The timing of this survey (in the summer of 1991) is especially interesting because it provides a benchmark for the conditions of rural life in Russia immediately prior to the collapse of the Soviet Union. This survey shows that even in 1991 there were substantial differences between villages and individuals with respect to household production and sales and subjective quality of life, as well as degree of support for what was to become an emerging market economy in agriculture and social services. Most important, this study includes some important indicators of human capital in the peasant household, as well as some indicators of community attachment, an important type of social capital.

A focus of the 1991 survey was the extent to which rural residents would support privatization of various types of social services. Since 1991 this has become a central issue in efforts to restructure Russian agriculture, as the post-Soviet large farm enterprises, the successors to the *kolkhozy* and *sovkhozy*, have been forced by market and potential forces to weaken their traditional supports for service provision (O'Brien et al. 1993). The 1991 survey identified specific ways in which peasant households were linked to the *kolkhozy* in the Soviet period, and how these attachments have affected their views toward economic and political change in Russia. In addition, the survey, along with a comparable survey from rural communities in Missouri, provides important comparisons of quality of life indicators with a sample of Mid-Western American rural households (Patsiorkovski and O'Brien 1996).

The 1993 survey also contained items pertaining to rural residents' service preferences, thus providing a basis of comparison between the end of the Soviet period and the beginning of the post-Soviet period. Most important this survey contained items that measured the social exchange networks of peasant households, thus providing an important measure of social capital (Dershem 1995 1998). In addition, this survey used the Center for Epidemiological Study of Depression Scale (CESD), which provided an important benchmark measure of peasant household mental health in the early years of the post-Soviet period (Dershem et al. 1996). The 1993 survey contained semantic differential scale measures of subjective quality of life (Campbell et al. 1981; O'Brien et al. 1998b:199-220). Indicators of change in household production and sales, income and acquisition of animals and durable material goods, such as automobiles and VCRs provided an important basis of comparison of emerging stratification and inequality between households from 1991 to 1993 (O'Brien et al. 1998b: 87-105).

Selection of Villages for the Panel Study

Figure 4.1 shows the location of the three villages in the panel study. Two of the villages, Latonovo in Rostov *Oblast* in the North Caucasus Region and Vengerovka in Belgorod *Oblast* in the Central Black-earth Region, in the black earth zone (*chernozem*), and one village, Bolshoe Sviattsovo in Tver' *Oblast* in the Central Region is outside of this in the *podzol* or non-black earth zone.

Rural areas of Russia are quite diverse in population, climate, soil, and other conditions that affect not only peasant household production and sales, but also the capacity of households to generate and maintain human and social capital (see Chapter 3). The area of Russia which historically has been the most productive agriculturally, both in pre-Soviet and Soviet times, has been in the *chernozem* zone of southern and south central Russia. This zone, which contains very fertile soil, and a more temperate climate than in northern Russia, extends throughout southern and portions of central Russia and contrasts with the *podzol* zone in northern Russia which has poor soil and a much colder climate.

64 Household Capital and the Agrarian Problem in Russia

Figure 4.1: Village Location

Because the *chernozem* zone is the most fertile for agricultural production, it was the site of the earliest and most vigorous efforts to develop collective farms. During the Soviet period, *kolkhozy* and *sovkhozy* in the *chernozem* zone were much larger and more administratively centralized than their counterparts in the *podzol* zone. The *kolkhozy* and *sovkhozy* in *chernozem* zone were highly capital intensive, specializing in various mixes of grain, beef and dairy production. By and large, these large enterprises continue to operate in a restructured form as *TOO*s. Regardless of the type of economic and political system that may evolve in the future, it is clear that this area will be continue to occupy a dominant role in Russian agricultural production. Because of this, the villages in the *chernozem* zone have a much greater potential to expand human and social capital than is the case in less agriculturally productive areas in the country.

The first village selected from the *chernozem* zone was Latonovo, in Rostov *Oblast*. It was included in the first joint Russian-American rural village survey that was conducted in 1991 (O'Brien et al. 1998b;

Patsiorkovski and O'Brien 1996). (Because of growing differences between Russian and Ukrainian economic policies, another village in the 1991 survey, Mayaki in Donets *Oblast* in Ukraine, was dropped from the joint project after 1993.) In 1996, the population of Latonovo was 1,509 (532 households). The village is located 745 miles south of Moscow in the Matveev-Kurgen Region on the southern edge of the Russian plains, in Rostov *Oblast*. By train, the trip from Moscow to Matveev-Kurgen, during which the Ukrainian border is crossed twice, takes 22 hours. The remaining 11 miles to Latonovo must be traveled by bus, taxi, or car. The road between Latonovo and Matveev-Kurgen is paved with two lanes of relatively good quality. Residents of Latonovo travel to Matveev-Kurgen, with a population of 7,000, to market produce and meat from their private plots, as well as for shopping and employment.

The nearest city is Taganrog, the regional administrative center, which has a population of slightly fewer than 300,000. It is located 31 miles from Latonovo. Taganrog is far enough away from Latonovo that it is unlikely that Latonovo will become a suburban "bedroom community." But, Taganrog is close enough to Latonovo that village residents are not isolated to the degree that is experienced in more remote villages in northern Russia. Residents of Latonovo go to Taganrog for marketing of produce and meat from their private plots and to shop, and a few villagers are employed there. For example, in 1991, 92.3 percent, and in 1997, 94.9 percent, of employed persons from Latonovo worked in their own village or in a nearby village. The majority worked for the *TOO*, while most of the remainder worked in village services, such as the post office, food and clothing shops, clinics, and the local Village Council (formerly *Sel'soviet*). In 1991, 0.6 percent of the labor force was employed in Taganrog, increasing to 3.4 percent in 1997. There was no change in the percentage of the labor force employed in the city of Rostov in 1991 and 1997 (1.6 and 1.7 percent respectively).

During the first stages of collectivization in the early 1930s, Latonovo was identified as the labor base for the *kolkhoz*, *Znamia Lenina* (Lenin's Banner). In 1992, the members of the *kolkhoz* changed the name to *Tovarishchestvo s Organichennoi Otvetstvennost'u (TOO) Zakrutogo Tipa Latonovo*, which roughly translates to "Private Joint-Stock Company or Association Closed Type of Latonovo." The *TOO* produces cereal grains, sunflower, and beef.

The village of Vengerovka, in Belgorod *Oblast*, with a population of 1,010 (267 households) in 1996, is located 500 miles south of Moscow, approximately 31 miles north of the Ukrainian border in south central Russia. The village is 48 miles from the city of Belgorod, which has a population of 311,000. The bus ride between Vengerovka and Belogorod takes slightly more than two hours. Thus, as in the case of Latonovo, it is unlikely that Vengerovka will become a suburban bedroom community, but villagers are not nearly as isolated as they would be if they were living in northern Russia.

Vengerovka was identified during the period of collectivization as the labor base for the *kolkhoz*, *Krupskaia* (Krupskaia was Lenin's wife). In 1992, the members of the *kolkhoz* changed the name to *Tovarishchestvo s Organichennoi Otvetstvennost'u zakrutogo tipa Vengerovka,* which roughly translates to "Private Joint-Stock Company or Association Closed Type of Vengerovka". As in Latonovo, the *TOO* produces cereal grains, sunflower, and beef.

In 1993, 92.1 percent of employed persons in Vengerovka worked in their own village or in a nearby village. This figure increased to 97.7 percent in 1997. The majority of working-age adults in Vengerovka, as in Latonovo, worked for the *TOO*, while most of the remainder worked in village services, such as the post office, food and clothing shops, clinics, and the local village council. In 1993, 6.6 percent of the labor force was employed in Belgorod, decreasing to 1.1 percent in 1997. There also was a decrease in the percentage of the labor force employed in the city of other large cities in 1993 and 1997 (2.3 and 1.1 percent respectively).

Although Latonovo and Vengerovka share much in common, in terms of quality of land, climate, distances to markets, and demographics, they are located in *oblasts* which differ markedly in their approaches to support for peasant households. This is seen especially in the way these *oblasts* support the development of human and social capital within those households. Belgorod *Oblast* offers peasant households access to a special fund that allows them to borrow money to improve existing homes or to build new homes and buildings for storing grain or silage, or for keeping animals. Rostov *Oblast* does not have such a fund for generating financial capital, but offers support for peasant household human and social capital development through informal mechanisms of kin and village networks (O'Brien et al. 1998a: 44-45).

The third village in the panel study is Bolshoe Sviattsovo ("greater Sviattsovo"), which is a collection of eight very small villages, in Tver *Oblast*. It is located in the *podzol* zone in northern Russia, 150 miles northwest of Moscow and 300 miles south of St. Petersburg. Bolshoe Sviattsovo is seven miles from Torzhok, with a population of 70,000 people, and 50 miles from Tver the capital of the *oblast*.

The population of Sviattsovo Bolshoe was 920 (241 households) in 1996. The climate in this village is much harsher and the soil much less productive than in Latonovo or Vengerovka. Although this village is at a comparative disadvantage vis-à-vis the two black earth zone villages with respect to agricultural productivity, its proximity to Moscow markets gives peasant households some offsetting advantages. In addition, the close proximity of the village to Torzhok provides at least the potential for a bedroom community relationship that is not found in the other villages.

The village reflects the approach to agricultural reform that is found in Tver *Oblast*. Tver *Oblast* has been less reform-minded than either Rostov or Belgorod *Oblasts* (O'Brien et al. 1998a: 42). This is reflected in the decision by the *kolkhozniky* in Bolshoe Sviattsovo in 1992 to remain a *kolkhoz*. The *kolkhoz* in Bolshoe Sviattsovo produces a variety of dairy products, some grain, some fiber (flax), and wood products.

In 1993, 95.5 percent of employed adults living in Bolshoe Sviattsovo worked in their own village or in a nearby village. This figure increased slightly to 96.4 percent in 1997. The majority of persons worked for the *kolkhoz*, with most of the remainder working in village services, such as the post office, food and clothing shops, clinics, and the local Village Council. In 1993, 3.0 percent of the labor force was employed in Torzhok, decreasing to 2.2 percent in 1997. There was no change in the percentage of the labor force employed in the city of Tver in 1993 and 1997 (1.5 and 1.4 percent respectively).

Sampling of Households and Respondents

One of the great benefits of Soviet era record-keeping practices, which still are found in many areas of Russian life, is that they permit a much simpler and much more accurate process for identifying samples than is the case in the United States or Western Europe. Detailed lists of household and individual characteristics in each village facilitate a sampling scheme that

68 *Household Capital and the Agrarian Problem in Russia*

maximizes the opportunity to examine variations in household labor and social capital.

The household sample was constructed from the official list of permanent residents in each village, which is called the *kniga ucheta domashnih khoziaistva* (Book of Household Accounts). The Village Council updates this list at the beginning of each calendar year. These very accurate records provide detailed demographic and social characteristics of all households within its jurisdiction.

Table 4.1 presents the number of households and individuals, sample size and percentage of households interviewed in each of the three villages from 1995 to 1997. In 1995, approximately 45 percent of all households in the three villages were interviewed. Due to a small out-migration in all of the three villages, the percentage of all households interviewed for the three years remained constant. In the initial year of the study, 1995, the lowest proportion of households interviewed was in Latonovo (29.6 percent) due to its larger size than the other villages. The percentage of households interviewed in Vengerovka and Sviattsovo was 59.5 and 62.8 percent.

Table 4.1: Number of Households and Individuals, Sample Size and the Percentage of Households Interviewed in the Three Villages from 1995 to 1997

Village	1995			1996			1997		
	Households (individuals)	Sample	%	Households (individuals)	Sample	%	Household (individuals)	Sample	%
Latonovo	531 (1540)	157	29.6	532 (1509)	157	29.5	534 (1598)	157	29.4
Vengerovka	262 (1002)	156	59.5	267 (1010)	156	58.4	261 (1020)	156	59.8
Sviattsovo	239 (927)	150	62.8	241 (920)	150	62.2	250 (916)	150	60.0
Total	1032	463	44.9	1041	463	44.5	1045	463	44.3

A stratified sample was obtained based on the proportion of seven different household types in each village. These household types have been used in studies of household production, consumption and service needs by the ISESP since the late 1980s (see Patsiorkovski 1991; Patsiorkovski and Rimashevskaya 1991). They were used in the joint Russian-American

Russian village surveys in 1991 and 1993 (O'Brien et al. 1998b; Patsiorkovski and O'Brien 1996). These seven types and their specific proportions in the villages in 1995 are:

(1) **single adult household** [23.2 percent]- includes people of various age groups, who live separately and keep an independent household;

(2) **retired couples** [9.3 percent]- contains a married couple in which the male is 60 years of age or older;

(3) **employed couples without children** [8.7 percent]- typically, this includes parents who are of employment age who are helping their children who have migrated to urban areas;

(4) **employed couples with children** [28.9 percent]- nuclear families;

(5) **employed couples with children and other adults** [10.0 percent]- nuclear families, usually with relatives such as parents of one of the spouses or a grandfather/mother, brother or sister;

(6) **single parents with children** [2.6 percent]- usually, a mother with a child under the age of 16;

(7) **other extended family** [17.3 percent]-primarily parents living with adult but unmarried children or grandmothers living with their grandchildren. Included in this family type are households comprised of a brother and a sister, or an aunt/uncle with a niece/nephew, or some other combination of adults living together.

The primary public policy concern of the ISESP during 1991 and 1993 was to identify household patterns of consumption and preferences with respect to services, as well as household production and sale of products from private plots. Therefore, interviewers were instructed to interview the *khozian,* or the "head of household" who is responsible for social services and plot production. This led to a sampling bias of greater selection of women, who were more apt to be in charge of this area (Benet 1970; Danes et al. 1994), than would have occurred if there was a random selection of adult household members as would be the case in American surveys. Nonetheless, given the higher life expectancy of Russian women, which is on the average 10 years greater than that of Russian men, the proportion of women in the sample is not that much higher than their proportional representation in the populations of the villages in the study.

Moreover, although the sampling procedures for selecting respondents resulted in some clear biases, the alternative of opting for a more traditional American random respondent selection procedure for the panel study would have created other, and from our view, more serious biases. Specifically, this would have made it very difficult to make accurate longitudinal comparisons between survey data gathered in 1991 and 1993 with survey data gathered in the three waves from 1995 to 1997. Therefore, the researchers decided to keep the respondent selection procedures employed in the 1991 and 1993 surveys.

The following is an overview of the demographic characteristics of the 463 panel respondents at the start of the panel study in 1995. Almost three-quarters of respondents were women (74.5 percent). Respondents were, on average, 52 years of age; 50 for men and 52 for women. Two-thirds (65.8 percent) of all respondents were married, 27.5 percent were widows(ers), 4.1 percent were divorced, and 2.6 percent were single (never married). The average number of years of education was eight, and almost one-third had eleven or more years of education.

More than two-thirds, 68.3 percent of the respondents were employed in the local *TOO* or *kolkhoz*, 25.8 percent worked in the social or public service, 2.3 percent were newly registered private farmers, 1.4 percent were involved in other agriculturally-related businesses, and 2.3 percent were engaged in nonagricultural-related businesses.

A total of 524 individuals were interviewed in the first wave of the panel study in 1995. Over the next two years (1996 and 1997), sixty-one of the original respondents had either moved, died or were unwilling to participate, resulting in a panel of 463 of the same respondents who were interviewed three times over the three-year time period. The resulting attrition rate was 11.6 percent.

Operationalization and Measurement

Indicators were developed for the following variables that are included in the hypotheses described in Chapter 2.

Agricultural Production and Sales

The measure of production and sales of agricultural products in the household was based on respondents' reports of the number of kilograms of specific agricultural products that their households actually produced and sold in the previous year. This included fruit, meat, milk, potatoes and vegetables, in both unprocessed and processed forms. To determine the

Research Design 71

value of these products, each product was weighted according to its ruble value in the nearest regional market during the time in which the surveys were conducted. In 1995, for example, potatoes and vegetables were sold for approximately 4,500 rubles, which was approximately equal to one US dollar at that time. Milk was sold for approximately 1.5 times the price of potatoes and vegetables, while meat was sold for approximately six times the price of potatoes and vegetables. These figures were summed to produce an overall weighted amount of kilograms of household production based on the value of products produced. This became the basis for measuring nonmonetized as well as monetized income, since the value of products produced and consumed by the family was an important part of the total income of the peasant household.

The specific measure of agricultural sales was the weighted ruble value, according to the weights described above. In 1995, 61.6 percent of the households in the sample had some agricultural sales, decreasing to 58.7 percent in 1996, and then increasing to 71.3 percent in 1997. Because a few very high producers, the two or three households in a village who were officially registered as private farmers, had sales which were dramatically higher than other families, a logarithmic transformation was employed (O'Brien et al. 1996b).

Material Goods

Respondents were asked during each wave of the panel study to report any purchases of major durable goods. This included purchases of cars, VCRs and telephones (which require very high outlays of money to have installed).

Physical Capital

Three types of physical capital in the household were measured: 1) the number of animals owned, 2) the amount of land rented from the Village council, and 3) mechanical equipment owned by the household.

The number of animals owned by the household included poultry, pigs and cows. A weighting scheme was used which took account of the amount of raw or processed products, which could be gained, from a single animal of a given type. The weights were 1 for poultry, 50 for a pig and 250 for a cow. These figures were summed and a logarithmic transformation was made to deal with the fact that a small number of households had a substantially larger number of poultry than other households.

There are three types of land tenure arrangements: private plots of land surrounding or adjacent to the household; communal land allotted to households by the local Village Council; and land rented from the Village Council. All three types of land tenure were measured, although primary focus was with variations in amount of rented land.

To gauge the amount of mechanical equipment owned by a household, respondents were asked if they had any mechanical equipment for agricultural production or processing. This included a range of items, such as cream separators, fruit tree spraying equipment, various kinds of cultivating equipment, and equipment for curing hams or sausages. The proportion of households having very expensive equipment, such as tractors, was only 1.6 percent, but the proportion of households having any type of mechanical equipment was 9.8 percent (O'Brien et al. 1996b).

Household Labor

The operationalization of household labor was based on a formula that is an adaptation from the earlier work of Chaianov (Chaianov 1966; Deere and de Janvry 1981). A weighted labor potential, based on age, was given to every household member. The weights were: 0 (less than 8 years of age, and 80 years and older), 0.25 (8 to 11 years, and 75 to 79 years), 0.50 (12 to 14 years, and 71 to 74 years), 0.75 (15 to 16 years, and 66 to 70 years), and 1 (17 to 65 years). These figures were then summed for each household. This created a measure of the amount of household labor in each household.

Social Networks and Community Attachment

Several types of social capital were measured. Social exchange helping networks were measured with items that were similar to the social network items used in Fischer (1982). These items were adapted to meet the specific social organizational, needs and cultural characteristics of the Russian village. These quantitative indicators were first developed for the 1993 survey and their validity was checked during intensive three month long ethnographic observations in the summer of 1993 (Dershem 1995). Quantitative and qualitative data from the summer of 1993 were then used to refine and expand the quantitative measures of social exchange networks which were utilized in the panel study.

Respondents were asked to identify up to five specific individuals who assisted the household in five specific areas: (1) borrowing money, (2) trading goods and services, (3) taking care of the household if someone is sick, (4) assisting with the cultivation of the household plot, (5) assisting with household tasks, and (6) discussing important matters. Two summary measures of network characteristics were created: redundant and nonredundant ties. Redundant exchange networks included the total number of persons identified in each helping area, regardless of whether or not a person was identified in another area (Fischer 1982:139), referred to as multistranded ties. Nonredundant exchange networks included only the sum total of unique individuals for all helping areas.

Community attachment was measured in two ways. First, by asking respondents the extent to which they attended both family and village events and ceremonies; (0) never, (1) sometimes and (2) often. These responses were summed for a total community involvement score. The alpha reliability of this index was 0.72 in 1995, 0.69 in 1996 and 0.70 in 1997.

Second, a subjective measure of community integration was obtained by asking the respondent how he or she "fit" into the local community. This index has been used in surveys in rural communities in the American Midwest and has been found to have a strong association with mental health outcomes (O'Brien et al. 1994). In this study, the index was comprised of two items. The first asked the respondent to respond on a seven-point scale to the question, "how well do people in the village relate to you?" The second question asked, "How much do you have in common with most of the people in your village?" These responses were summed for the sense-of-fit score. The alpha reliability of this index was 0.68 in 1995, 0.67 in 1996 and 0.67 in 1997.

Depressive Symptoms

The specific indicator measuring depressive symptoms was an eleven-item modified version of the Center for Epidemiological Study of Depression Scale (CES-D) (Mirowsky and Ross 1989; Radloff 1977) which had been used in the 1993 study (Dershem 1996; O'Brien et al. 1998b: 199-220). Confirmatory factor analysis of the 1993 data showed that although CES-D scores are quite higher in Russian than in American rural samples, the factor structures are very similar in both samples (Dershem et al. 1996). The factor analysis of the CES-D for each year of the panel study is presented in Appendix 12. The alpha reliability of the CES-D scale in the Russian sample was 0.73, 0.85 and 0.88 in the 1995, 1996 and 1997 surveys respectively.

To control for other factors which might affect levels of depressive symptoms, variables which earlier studies had shown to be associated with mental health also were included in the analysis: age, education, gender, health satisfaction, and negative life events (Ensel 1986; O'Brien et al. 1996).

Subjective Quality of Life

The standard type of seven-point semantic differential scales developed by Campbell and his associates (Campbell et al, 1976), and used in the 1993 survey (O'Brien et al. 1998b), were used to measure subjective quality of life in seven life domains and life-as-a-whole. Respondents were asked to respond to three pairs of 7-point scale semantic differential items - dissatisfied...satisfied, unpleasant...pleasant, disappointing...rewarding – to rate their level of satisfaction with each of the life domains and their life-as-a-whole. The seven domains included (1) employment (not analyzed), (2) income, (3) marital status, (4) family relations, (5) village life, (6) the situation in the country, and (7) health satisfaction. Satisfaction with health was obtained by using one 7-point semantic differential scale –dissatisfied to satisfied (McClendon and O'Brien 1988). The alpha reliabilities of the indexes created by combining the three semantic differential items for the six life domains and life-as-a-whole are shown in Table 4.2.

Table 4.2: Cronbach Alpha Reliabilities for Various Life Domains and Life-as-a-Whole Indexes for 1995, 1996 and 1997

Life Domains	1995	1996	1997
1. Income	0.93	0.96	0.93
2. Marital status	0.96	0.98	0.96
3. Family relations	0.97	0.97	0.96
4. Village life	0.96	0.97	0.94
5. Situation in country	0.92	0.97	0.91
6. Health*	n/a	n/a	n/a
7. Life-as-a-whole	0.96	0.93	0.93

* Health satisfaction was measured by one variable.

Interviewing and Validation of Survey Questions

Construction of interview questions and plans for pre-testing and interviewing were conducted jointly by the American and Russian research teams in Columbia, Missouri and Moscow. Interviewing was preceded by several visits to the villages, in the winter, by the Russian principal investigator, during which time he secured the support of local village administrators, *TOO*, or *kolkhoz* chairmen, and other significant officials for the project.

Full-time researchers from ISESP, who had worked in survey research settings for many years, conducted the actual interviews with respondents in the three villages. The interviews were structured and face-to-face, taking over two-weeks in each village for each wave of the panel study.

In addition to structured formal surveys, 64 in-depth ethnographic interviews were completed with village residents, including village administrators, chairpersons of *kolkhozy* and *TOO*s, *fermery*, teachers, new entrepreneurs, pensioners and *krest'ianskie khoziaistva* over the three-year time period. These in-depth interviews provide a validity check on questions, as well as insights into the distributions of responses.

Statistical Methods and Strategies

The data analyzed in this study are from the same 463 respondents from the same households, for three consecutive years (1995, 1996, and 1997).

There are two basic strategies for testing the effects of variables over time with panel data. The first is a lagged cross-sectional (Kessler and Greenberg 1981; Tuma and Hannan 1984), which tests the effect of variables in one year on variables in each of the subsequent years.

The second, and the one used in this analysis, is a pooled cross-sectional strategy. The database of 463 cases for each of the three years is inverted to represent 1389 separate cases (463 cases x 3 years) which allows for the testing of the "pure" effect of a variable on another variable for the three year period of time. For example, instead of three cross-sectional data sets on the amount of household labor, there is one "pooled" data set representing 1389 cases on the amount of household labor. Pooled cross-section analysis is a type of generalized least squares (GLS) estimation in which observations are weighted to diminish the effects of serial correlation, thus protecting the integrity of the estimates. Interpretation of GLS estimates essentially are the same as those for OLS, indicating the main effect of an exogenous (independent) factor on the endogenous (dependent) factor, which shows the persistent relationships among variables in the model. Also, dummy variables for each of the data collection points (each year) test for changes in these relationships over time. This strategy will be used to test the theoretical models shown in Chapter 2.

The initial wave of data (1995) was entered into Version 6.1 of the SPSS Windows program. The second wave of data (1996) was entered, and the 1995 database was converted to, Version 7.5 of SPSS Windows. In the final year (1997) data were entered into, and the previous database was upgraded to, Version 8.0 of SPSS. Data entry and cleaning was performed at the ISESP in Moscow. Data analyses were conducted at the University of Missouri in Columbia, Missouri using Version 8.0 of SPSS.

All structural model estimations were performed in a single step using the Amos Version 3.61 Program (Arbuckle 1997), supported by SPSS. Structural modeling was used for several reasons. First, because the relationships studied in this research project are complex, models must simultaneously handle a series of independent (exogenous) variables and a series of dependent (endogenous) variables and their interdependencies. Moreover, causal flows need not be one way. Second, the models must accurately reflect complex relationships because any variable, observed or latent, can predict any other variable. Structural modeling produces a more comprehensive model because all of the equations in the model are solved

Research Design 77

simultaneously. Third, in panel data, measurement error and autocorrelated errors associated with two or more of the same indicators can influence estimates. Structural modeling considers measurement error so that these estimates are less biased than Ordinary Least Squares regression estimates.

Summary

The basic design of this study is a micro-level analysis of the adaptation of households to an emerging market economy in rural Russia over a three-year period of time. Data were obtained from a proportional sample of 463 households in three villages geographically dispersed throughout European Russia; one village located in the northern (Tver *Oblast*), central (Belgorod *Oblast*) and southern (Rostov *Oblast*) areas. Multiple indicators, pretested in previous studies of households in these villages, were used for all measures.

Two approaches were used in the data analyses. First, bivariate descriptive statistics were used to show basic relationships between household capital (household labor, community attachment and social networks) and agricultural production, agricultural sales, symptoms of depression and evaluations of various quality of life domains. Second, structural equation models were used to present comprehensive analyses of the multiple relationships simultaneously.

5 Household Structure and Labor

David J. O'Brien, Larry D. Dershem and Valeri V. Patsiorkovski

Introduction

In this chapter we will explore the question, what kinds of human capital are available in Russian rural households that are associated with greater or lesser economic and psychological adaptation for their members? Our attention will be directed to the differential ability of villages and households to aquire the household hand-labor that is an essential part of the emerging institution of private farming in Russia.

Our macro-level analysis in Chapter 3 showed that there is enough human capital, in terms of able-bodied adults, in the major agricultural producing regions of rural Russia, to support a significant expansion of private agricultural production, if appropriate land relations and other conditions are developed. However, the quantity and quality of this human capital varies considerably from one region of the country to another. The number and age structure of households in the *chernozem* zone, which includes the Central Black-earth and North Caucasus Regions, in southern and south-central Russia, for example, provide much more favorable conditions for the expansion of private agriculture than do the corresponding numbers and age structures of households in the *podzol* zone of Central and Northern Regions of Russia. Vengerovka in Belgorod *Oblast* is located in the Central Black-earth Region and Latonovo in Rostov *Oblast*, which is located in the North Caucasus Region, while Boshoe Sviattsovo in Tver *Oblast* is located in the Central Region.

The sampling design of the panel study provides us with an opportunity to examine at the micro-level how household structure, which is a form of social capital, affects the ability of a household to secure human capital, in the form of household labor. In subsequent chapters we will examine how the differentiation of Russian rural households in household labor is reflected in an increasing differentiation in the ability of individual households to compete in the new market economy.

80 *Household Capital and the Agrarian Problem in Russia*

Household structure, which refers to the relationship (or lack of relationships) between various types of household members (husbands, wives, children and other relatives), is the social glue that holds together whatever human capital the household is able to muster. *Different types of social structures provide relative advantages or disadvantages, insofar as they provide the social capital within which the household's capacity for hand labor is embedded.* Specifically, some households are better able than others to mobilize *cooperative* efforts involving hand labor that contributes to household collective goods, in the form of agricultural production and sales.

A Demographic Profile of the Households in the Sample Villages

The distribution of demographic types of households in the panel study is shown in Table 5.1. A description of these types and the stratified random sampling procedure that was used to locate them for the sample are found in Chapter 4. Households, proportional to their actual number, were randomly drawn from each demographic household type within each village.

Over a third of the sample (34.1 percent) in 1995 consists of households that are structurally incapable of having substantial human capital to assist in household production; single persons (largely elderly widows), retired couples and single parents. The remaining, almost two-thirds, of the households (66.4 percent) in 1995, however, are structurally set up to produce fairly high levels of human capital (see Chapter 3).

Table 5.1: Demographic Types of Households in the Panel Study Sample from 1995 to 1997

Demogrpahic Types of Households	1995	1996	1997
1. Single person	21.6	20.7	21.8
2. Retired couple	9.9	10.6	11.2
3. Employed couple	8.2	7.1	6.9
4. Employed couple with children	30.7	31.3	30.2
5. Employed couple with children and other adults	9.9	10.8	12.8
6. Single parents	2.6	3.9	3.2
7. Other	17.1	15.6	13.8
Total	100.0	100.0	100.0

During the three-year period of the panel study, the structures of some households remained fairly stable, while there were changes in some other households. The proportion of households classified as "single person" remained virtually the same during the three-year period. This type consists largely of elderly widows. The proportion of households containing an employed couple with children also remained virtually unchanged. The proportion of households listed as "retired couples" increased, due simply to the aging of the panel sample. There is also a fairly strong increase in the number of single parent households from the first to the second year of the study, but this percentage begins to decline in the third year of the study. The proportion of single parent households remains a very small proportion of the total village population.

The most interesting changes are those involving the third, fifth and seventh types of households; "employed couples without children," "employed couples with children and other adults" and the "other" category. Types three and seven are somewhat unstable, insofar as households in these categories are likely to move into other categories. Type three, "employed couples without children," is a very heterogeneous group that includes middle-aged adults whose children have just left the household for university training or employment in some other location, as well as older couples who are close to retirement age. Because of this heterogeneity the human and social capital of households in this type varies considerably and, in turn, so does their economic productivity. We will see in Chapter 6, for example, that an important type of social capital, social networks, varies considerably among households in type three. Because of these differences, there is considerable variation in the propensity of type three households to rent land.

When older couples in type three retire, their households will be defined as type two, "retired couple" households. Alternatively, a younger couple without dependent children at home would be re-defined as demographic type seven, "other," if its adult son returned to the village with his wife and children and lived with them.

Type seven, the "other" category, refers to households that have been constituted in some way that does not follow traditional rural Russian village norms. This residual category expanded very rapidly during the early years of economic crisis, from 1992 to 1995, in which many urban Russians found themselves without employment and elected to move to the countryside to live with relatives. This was also a period in which there was an increase in the number of Russians from other Soviet Republics in the "near abroad" who were forced to find a place to live (Wegren 1998:22-24).

These households, however, were created in response to an emergency and thus are less stable structures to maintain human capital than the households that fall into type five. Thus, for example, when economic conditions improve or refugees from other former Soviet republics get oriented to their new situation, there is a corresponding out-migration of persons from the seventh type of household. This is reflected in the substantial decline in the number of households in this demographic type from 17.1 to 13.8 percent in three years.

This situation might include, for example, a fifty-year old nephew of an elderly widow who was displaced from a defense industry factory'job in the city of Belgorod and moved back temporarily to stay with his aunt in a peasant household in Vengerovka where he could at least feed himself. This move typically represented a short-term survival strategy that was a response to extreme economic dislocation during that time period. When, however, the nephew left the household to take a job in a city, the household reverted back to type one, a single person household. In another situation, an adult child, with a wife and children, moved back to his parents home during bad economic times, thereby changing that household from type three, "adult couple without children," to type seven, "other." When, however, the adult child and his wife and children moved out of his parents'·home, to take advantage of new economic opportunities in the city, his parents' household type would return to "adult couple without children."

Alternatively, type five, "employed couples with children and other adults" represents relatively stable family structures that contain members from two or three generations. The substantial growth in the traditional multi-generation extended family type is an indication of a more significant and longer-term restructuring of rural households as more and more young people and extended family members in the villages have found an economic niche in household agricultural production.

Bolshoe Sviattsovo, the northern village that is outside of the *chernozem* zone, has a larger number of single person households (28.3 percent in 1997), made up largely of elderly widows, than do either Latonovo (15.9 percent in 1997) or Vengerovka (21.3 percent). Alternatively, Bolshoe Sviattsovo has a much smaller proportion of

Household Structure and Labor 83

households in type five, employed couples with children and other adults (5.3 percent in 1997), than Latonovo (15.3 percent in 1997) and Vengerovka (17.3 percent in 1997). This illustrates that Bolshoe Sviattsovo, which is located in the *podzol* zone, is much less sustainable as an agricultural community than the two villages located in the *chernozem* zone. (Detailed information on the distribution of different demographic types of households in the three villages in the panel study is presented in Appendix 3).

Table 5.2: Percentage of Households Which Decreased, Increased or Had No Change in Size from 1995 to 1997

	1995-1996	1996-1997	1995-1997
Decreased size	6.7	6.5	10.4
No change in size	86.2	87.9	78.9
Increased size	7.0	5.6	9.7
Total	100.0	100.0	100.0

Table 5.2 provides a summary of the reconstitution of households during the three-year period of the panel study. Approximately the same number of households increased as decreased in size from 1995 to 1997. The primary cause for a decrease in household size was the death of household members and movement of children out of the household because of marriage or completing school. The primary reason for increasing household size was the addition of new members to intergenerational extended families when children moved back home or an adult couple moved in with their adult children to consolidate a household production operation. In other instances an elderly person in her nineties died and was replaced in a household by a young couple with children or a divorced man brought a new wife and her children to his household in the village.

The total change in the sample households, including both increases and decreases, during the relatively brief, three year, period of the panel study was more than twenty percent. Given the highly conservative nature of traditional family structures in rural Russian villages—see, for example, the low divorce statistics cited below—this is a rather substantial change for such a short period of time!

84 *Household Capital and the Agrarian Problem in Russia*

In Latonovo and Vengerovka, the percentage of households in the type with the highest level of human capital, employed couples with children and other adults, expanded during the course of the study, from 10.8 to 15.4 and from 14.7 to 17.3, respectively, from 1995 to 1997 (see Appendix 3). In Bolshoe Sviattsovo, however, the percentage of households of this type increased only slightly, from 4.0 to 5.3, during the three years.

One other important difference between the villages is in the number of single parent households. The percentage of households of this type (type 6) in Latonovo is 5.1 in 1995, 8.3 in 1996 and 7.0 in 1997. This contrasts with much lower figures in Vengerovka (1.3, 1.3 and 0.6) and Bolshoe Sviattsovo (1.3, 2.0 and 2.0) for the same years. Ironically, the higher figures for this type of household in Latonovo than in the other two villages reflects the much stronger extended kinship networks in Latonovo that provide spouses who are abused and/or living with alcoholic husbands with a place to live that permits them to file for divorce. In the other villages, women are more likely to stay in difficult situations simply because if they leave their husbands they will be unable to find a place to live [see Chapter 6 and Dershem (1995) for an expanded treatment of the different social network characteristics of the three villages].

Table 5.3: Age Structure of All Household Members by Gender and Year

	1995		1996		1997	
Age groups	Males (n=605)	Female (n=731)	Male (n=600)	Female (n=738)	Male (n=603)	Female (n=726)
<5	6.1	5.2	5.5	3.9	3.8	3.9
5 – 17	25.8	20.0	26.3	21.5	26.5	20.9
18 – 24	6.4	5.6	5.8	4.9	6.6	4.8
25 – 44	32.6	26.4	31.2	26.7	31.2	26.7
45 – 64	20.5	20.2	21.8	19.8	20.9	19.1
65 – 74	7.4	13.8	8.2	14.2	9.3	14.6
>74	1.2	8.8	1.2	8.9	1.7	10.0
Mean	32.9	40.4	33.6	40.9	34.4	41.8

Table 5.3 shows that the age structure of the households in the panel sample is very similar to the overall age structure of rural populations in Russia (Chapter 3, Table 3.4). There is a relatively high mean age in all households, with women having a substantially higher mean age than men because of the much higher life expectancy among the former. Nonetheless,

39 percent of the persons in the households in the sample in 1995 were in the prime ages of 18 to 44, which is clearly a sufficient number to support the development of a "farmer class" in rural Russian villages. These figures are very similar to what has been found in agriculturally dependent counties in the American Midwest. In one study of two agriculturally dependent communities in Missouri, for example, 32 percent of the persons in the sampled households were in the 18 to 44 year-old age group (O'Brien et al. 1998b:76). The major exception is that there is a smaller proportion of elderly men in Russian than in American rural areas, owing to the much larger gap in life expectancy between men and women in Russia than in the United States (Patsiorkovski and O'Brien 1996:136).

Not surprisingly, given the distribution of household types that was outlined earlier, the mean age of persons in households was higher in Bolshoe Sviattsovo than in Latonovo or in Vengerovka. In 1997, for example, the mean difference in the age of males in Latonovo and Bolshoe Sviattsovo was over 3 and one-half years (33.9 compared to 37.5), while the mean difference in the age of women in the two villages was even greater; almost 5 years (39.4 compared to 44.3). There is a substantial difference between Vengerovka and Sviattsovo in the mean age of males in 1997 (32.5 compared to 37.5), and a somewhat smaller difference between the two villages in the mean age of females in that year (42.1 compared to 44.3).

Table 5.4: Marital Status of All Adults 18 Years of Age or Older in Respondents' Households in 1995, 1996 and 1997 (in percent)

	1995		1996		1997	
	Men (n=412)	Women (n=547)	Men (n=600)	Women (n=738)	Men (n=603)	Women (n=727)
Married	82.3	62.3	83.2	61.9	81.0	62.5
Single	11.2	4.8	10.4	5.1	11.5	4.6
Widow/er	3.9	29.3	3.7	29.2	4.3	29.4
Divorced	2.7	3.7	2.7	3.8	3.1	3.5
Total	100.0	100.0	100.0	100.0	100.0	100.0

As presented in Table 5.4, the structure of the vast majority of households in the Russian villages contains social capital that is conducive to the growth of human capital. Typically, human capital is embedded in households in which there is a stable husband-wife relationship. Over 80 percent of the men in the sample live in married couple households, while almost two-thirds of the women live in this type of household. The largest percentage of single adult households consists of widows. The proportion of persons in the sample who are divorced is quite small, below four percent. This figure is consistent with the relatively lower proportion of divorced persons in rural than in urban Russia. In 1996, the number of divorced persons in Russia was 3.8 per thousand, with 4.4 per thousand in urban areas but only 2.2 per thousand in rural areas (Goskomstat 1997:128).

Consistent with our earlier observation about the greater number of single parent households in Latonovo than in the other two villages, there is a higher percentage of divorced persons living in Latonovo. In 1995, for example, the proportion of men living in Latonovo who were divorced was 5.4 percent, compared to 3.5 percent and 1.8 percent, respectively, in Vengerovka and Bolshoe Sviattsovo.

Table 5.5: Education Level of All Adults 18 Years of Age or Older in Respondents' Households in 1995 (in percent)

Educational Groups	Men (n=412)	Women (n=547)	Total (N=959)
< 10 years	41.0	53.9	48.4
10 years	31.6	15.2	22.2
> 10 years	27.4	30.9	29.4
Total	100.0	100.0	100.0

The households in the sample are, by world standards, quite well educated (see Table 5.5). Slightly more than half of the adult men and slightly less than half of the adult women have completed more than ten years of formal schooling (11 years is considered full primary and secondary schooling in Russia today).

There are, nonetheless, some differences between the amount of education in households in the three different villages. In 1995, a higher percentage of men and women in Bolshoe Sviattsovo had less than 10 years of formal education (34.6 percent and 44.3 percent) than in Latonovo (27.7

Household Structure and Labor 87

percent and 38.0 percent) and Vengerovka (23.1 percent and 39.4 percent). These differences reflect the older population in Bolshoe Sviattsovo.

Table 5.6: Average Number of Years of Education by Position for All Working Age Men and Women in Respondents' Households in 1995 (in percent)

Occupational Positions	N	Mean Years of Education
Leadership/managers	34	13.1
Specialists	69	13.6
Clerical	57	12.2
Collective farmers/ workers	369	9.5
Private farmers	12	11.9
Other	2	4.5
Total	533	10.6

Table 5.6 shows, as would be expected, that there is considerable variation in educational levels by occupation in the villages, with persons in leadership/managerial or technical occupations having at least some university training. Nonetheless, even among collective farmers, the least skilled job classification, the mean educational level is close to ten years.

The Labor Potential of Different Households in the Villages

The weighted labor potential for each household in the sample was calculated according to the criteria described in Chapter 4. For each household a summary labor potential score was obtained, based on the following weights for each of its members: less than 8 years of age and 80 or more years of age = 0; 8 to 11 years and 75 to 79 years = 0.25; 12 to 14 years and 71 to 74 years = 0.50; 15 to 16 years and 66 to 70 years = 0.75; and 17 to 65 years = 1.

Table 5.7: Distribution of Household Labor in the Total Sample for Each Year of the Panel Study

Household labor	1995 %	1995 n	1996 %	1996 n	1997 %	1997 n
0-1.74	30.9	143	31.5	146	31.7	147
1.75-2.74	35.4	164	38.2	177	37.4	173
2.75-3.74	23.3	108	20.5	95	20.1	93
3.75-4.74	7.3	34	7.6	35	9.5	44
4.75 +	3.0	14	2.2	10	1.4	6
Overall mean/N	2.1	463	2.1	463	2.1	463

The distribution of household labor in the total sample for each of the three years is shown in Table 5.7. For simplicity of presentation, the individual household scores have been grouped into five categories. In the structural equation modeling that follows in later chapters, the specific score for each household will be used. The mean level of household labor in the total sample remained the same during the three years of the panel study, but there was considerable movement of individual households from one level to another. There were gains in the number of households with the first (0-1.74) second (1.75-2.74) and fourth (3.74-4.74) levels and reductions in the number of households in the third (2.75-3.74) and fifth (4.75 +) levels. Because of these changes, subsequent analyses, in this and later chapters, will show differences in n-sizes for different household labor categories in different years.

Table 5.8 shows that the weighting system for household labor is consistent with our intuitive understanding of the relative capacity for human capital that is found in the different demographic types of households. Most of the single person households in the sample are made up of elderly widows; hence, the mean score is less than 1. Just being a couple, types two and three, increases household labor by a factor of three. Single parent households have slightly less human capital potential than households do in types two and three, even though single parent households, type six, contain a younger adult. All of the demographic types of households just described are relatively disadvantaged in human labor potential.

Table 5.8: Distribution of Household Labor by Demographic Type of Household in 1995 and 1997

Demographic Type of Household	Mean 1995	Mean 1997	Stnd. Dev. 1995	Stnd. Dev. 1997
1. Single person	0.57 (n=100)	0.52 (n=101)	0.38	0.37
2. Retired couple	1.73 (n=46)	1.64 (n=52)	0.31	0.38
3. Employed couple	1.93 (n=38)	1.95 (n=32)	0.17	0.19
4. Employed couple with children	2.62 (n=142)	2.64 (n=140)	0.62	0.56
5. Employed couple with children & other adults	3.78 (n=46)	3.79 (n=59)	0.92	0.73
6. Single parents	1.58 (n=12)	1.97 (n=15)	0.66	0.76
7. Other	2.32 (n=79)	2.14 (n=64)	1.05	0.81
Total	2.07 (n=463)	2.08 (n=463)	1.16	1.16

The most advantaged households are those in demographic types four, five and seven, although there are clear differences between each of these types, both in mean household labor potential and in the range of individual household capacity within each type. Being an employed couple with children, demographic type four, means that the household is made up of younger adults, and this is reflected in an increase of approximately 30 percent in household labor potential. Type five, employed couple with children and other adults, is clearly the most advantaged type of household with respect to the ability to access labor for household agricultural production. On average, this type of household has over one and one-half the labor potential of households that only contain an employed couple with children. Moreover, the standard deviation and range for this type of household is second only to the relatively unstable type seven.

Households in type seven have a mean household labor potential that is slightly less than households consisting of employed couples with children. At the same time, however, the standard deviation for type seven is quite large, indicating that within this type there is a considerable range of labor potential, depending on the ages of the persons who have been added to the household.

During the three years of the panel study there was a restructuring in many of the households in the sample that affected their levels of household labor. In 1995, employed couples with children accounted for 40.3 percent of the third level of weighted number of adults, from 2.75-

90 *Household Capital and the Agrarian Problem in Russia*

3.74, but by 1997 this demographic type only accounted for 28.0 percent of households with household labor from 2.75-3.74.

The most important changes during the three years occurred in the second highest category of household labor, 3.75-4.74. These are households that have a substantial potential for increasing their agricultural productivity and sales and, therefore, have a high potential for achieving a good quality of life for their members. In 1995, 21.4 percent of these households are from the demographic type defined as employed couples with children and other adults (type five). In the same year, 66.0 percent of the households with household labor from 3.75-4.74 were defined as the demographic type seven. By 1997, however, employed couples with children and other adults (type five) accounted for 44.3 percent of this category of household labor while the demographic type seven only accounted for 13.3 percent of this household labor category. This indicates that over time better quality household labor became concentrated in stable nuclear families.

Table 5.9 shows the distribution of household labor by village from 1995 to 1997. Consistent with our earlier findings on demographic types of households, Bolshoe Sviattsovo contains less household labor potential than the other two villages. This holds true for all three years of the panel study.

Table 5.9: Mean Distribution of Household Labor by Village for Each Year of the Panel Study

Villages	1995		1996		1997	
	Mean	Std. Dev.	Mean	Std. Dev.	Mean	Std. Dev.
Latonovo (n=157)	2.22	1.08	2.28	1.11	2.27	1.12
Vengerovka (n=156)	2.14	1.20	2.13	1.18	2.12	1.18
Sviattsovo (n=150)	1.18	1.06	1.82	1.07	1.92	1.13
Total (N=463)	2.06	1.13	2.08	1.13	2.07	1.16

Household Labor by Village for 1995: $F(2), 6.19, p < .001$. Household Labor by Village for 1996: $F(2), 6.75, p < .001$. Household Labor by Village for 1997: $F(2), 5.97, p < .01$.

Household Structure and Labor 91

Summary

In this chapter we have provided an overview of the human capital available in the three Russian villages in this study. By world standards, the overall educational level of the sample is quite high and thus conventional Western economic analyses would define these households as possessing high levels of human capital. Moreover, there is a higher percentage of persons in these villages who are in the prime 18-44 age-group than comparable rural communities in the American Midwest; 39 percent compared to 32 percent. In short, the basic demographic structure of the Russian villages provides ample human capital to take advantage of the opportunities offered by an emerging market economy.

A crucial source of differentiation between households is found in the potential they possess for household hand labor. As noted in Chapter 2, in the peasant household economy it is this capacity of household members to contribute to collective household production efforts that will affect their ability to produce and sell agricultural products from the household's plots. In turn this production and sales will be a major determinant of the household's economic and psychological well being.

We have proposed that this potential for household labor is strongly correlated with the social structure of the household. Thus, the demographic type of household, which is a form of social capital, forms the social capital context within which human capital in the peasant household economy is embedded. Most important, we have shown that our operational definition of household labor, which is theoretically linked to Chaianov's classical position (see Chapter 2), and operationalized in Chapter 4, is strongly associated with the demographic types of households that provided the basis for the sampling frame for the panel study.

The weighting procedure for identifying household labor potential is the most useful measure of human capital in structural equation modeling that will be shown in subsequent chapters because it permits us to adjust for restructuring within the households in the sample. This occurs when members leave, such as when children go outside the village to attend school, or when new members join the household, such as when adult children return to the household with their spouses and children and join their parents in a shared family agricultural enterprise.

The strongest potential for household labor is found in demographic type five households that are comprised of couples, their children and other adults. These households showed the greatest gain in household labor potential during the three years of the panel study. On the other hand, demographic type seven households, the "other" category, appear to have had a much more ephemeral labor potential, showing high levels of household labor in 1995 but losing much of that potential by 1997. *These differences between demographic types five and seven are ample evidence that although household labor potential is empirically defined as human capital, its stability or instability is derived from the social capital context in which it is embedded.*

Finally, our data show that there are some important differences between the three villages with respect to household labor potential. Not surprisingly, the village outside of the *chernozem* zone, Bolshoe Sviattsovo, has a somewhat older population than do the other two villages that are located in the *chernozem* zone. Moreover, the number of households in the demographic type with the highest household labor potential, couples with children and other adults, increased more in Latonovo and Vengerovka than in Bolshoe Sviattsovo.

We will now turn to additional factors that differentiate between households and villages with respect to the ability to produce and sell agricultural products. These are the two "purer" forms of social capital referred to in Chapter 2, community attachment and social exchange helping networks.

6 Community Attachment and Social Networks

Larry D. Dershem, David J. O'Brien and Valeri V. Patsiorkovski

Introduction

In this Chapter, we will examine three "purer" forms of social capital that affect the economic and psychological adaptation of Russian peasant households; *community involvement, sense of "fit" in the community and social exchange helping networks*. There is a substantial body of literature on these types of social capital in developed Western nations. One of the key findings reported in this literature is that rural residents relate to their communities differently and possess different kinds of personal helping networks than their urban counterparts (Kadushin 1983; Beggs et al. 1996).

Community attachment has a greater impact on the mental health of rural than urban residents in America (O'Brien et al. 1994). In addition, the personal helping networks of rural residents are much more "dense" (i.e., more persons know one another well) and contain more kin than the networks of urban residents (Fischer 1982; Kadushin 1983; O'Brien et al. 1996c; Beggs et al. 1996). This comparative data provides a very useful benchmark with which to evaluate the actual, as well as potential social capital that affects the ability of Russian rural households to deal with the exigencies of a market economy. At the same time, however, the specific nature of peasant household production, especially its labor intensive quality, described earlier, means that certain aspects of community attachment (community involvement and sense of fit) and social networks that are important in Western nations may not be as important in the rural Russian context today.

First, community attachment has somewhat different meanings and consequences in American and Russian rural settings. The economic costs of not being attached to the local community are much less severe in the American context of easy transportation and work alternatives than in the Russian context of generally poor transportation services and limited

94 Household Capital and the Agrarian Problem in Russia

economic opportunities outside of the local village (Patsiorkovski et al. 1991).

Community attachment affects subjective quality of life and symptoms of stress in both American and Russian rural contexts (Dershem et al. 1996; O'Brien et al. 1994; O'Brien et al. 1996b). But, poorer transportation and communication in rural Russia puts a greater premium on being connected to the local area than would be the case in the typical American rural setting. Even the most remote American village is not that far removed from the interstate highway system. This transportation system coupled with an efficient and relatively inexpensive telephone and internet access system means that individuals can receive help from persons in distant locations and thus it may not be as necessary to be connected to the local area as it would be in the Russian village.

Second, the types of personal social network ties that are needed in American and Russian rural areas are different. An important area of research in Western social network analysis is the "strength of weak ties." Such ties, that are based on casual acquaintanceship and require only a minimal amount of investment and obligation by interacting parties are very useful in providing information for professional job-seekers who wish to move from one company to another (Granovetter 1973). This type of relationship, however, is of little help in obtaining the very difficult and time-consuming labor necessary to develop and maintain the garden plot on a peasant household. Moreover, although trusting relationships involving the trade of labor and machinery are a fundamental part of American rural life, such relationships are more important in Russian rural life where almost all transactions take place in an informal economy that is outside of any kind of legal enforcement structure.

Demographic Factors Affecting Community Attachment and Social Networks

There are differences between the villages in our sample with respect to whether or not respondents have lived there since birth (see Table 6.1). Over two-thirds of the sample in Latonovo were born in the village, whereas the figure for Bolshoe Sviattsovo is only one-quarter. Vengerovka falls almost exactly in the middle between the two other villages on this score.

Community Attachment and Social Networks 95

Table 6.1: Respondents' Connections to Their Villages and the Local Area Through Birth and Kinship in 1995

Connections to the Village	Latonovo n=157	Vengerovka n=156	Sviattsovo n=150	Total Sample N=463
% Respondents born in village	71.1	53.2	25.3	49.5
% Respondents born in local area	79.6	85.9	62.6	76.2
% Respondents with an adult relative in village	88.5	81.4	88.0	85.8
Mean number of adult relatives in village	7.7	6.3	5.3	6.4

Scheffe test of differences between Latonovo and Bolshoe Sviattsovo in mean number of respondents born in their village: 1995, p<.05; 1996, p<.001; & 1997, p <.001. Scheffe test of differences between Latonovo and Vengerovka for 1996, p<.05.

A significant portion of those who were not born in their village, however, moved there from other villages that are located in their local area. During the Soviet period, it was common for villages perceived to be "inefficient" to be eliminated by the central government and for families in the destroyed villages to re-locate in nearby villages. This situation occurred frequently in northern Russia and is reflected in the higher number of persons in Bolshoe Sviattsovo who report that they were not born in that village but moved there from another village in the same local area. In two rural Missouri villages that were studied by the researchers in 1991, the proportion of respondents who reported that they were born in their village or in a nearby village was 68.8 percent.

The average length of residence of respondents in the three villages follows the trend just described, with a mean number of 41.6 years in Latonovo compared to 40.4 in Vengerovka and 33.6 in Sviattsovo. The average length of residence of respondents in the two rural Missouri communities studied by the researchers in 1991, was 33.6 years. Length of residence has been shown in studies in the United States to be positively associated with community attachment (Goudy 1990; Kasarda and Janowitz 1974).

Over 85 percent of the respondents in the total sample have at least one relative outside of the immediate household living in the same village. As would be expected, respondents in Latonovo have the highest average number of relatives living in their village (7.7), followed next by Vengerovka (6.3) and then by Bolshoe Sviattsovo (5.3). There is a statistically significant difference between the average number of relatives in the village in Latonovo and Bolshoe Sviattsovo during the panel study and between Latonovo and Vengerovka for 1996 (see Table 6.1). Nonetheless, even in the village with the weakest social networks, Bolshoe Sviattsovo, respondents still have an average of over five relatives outside of their immediate household who are located in the village.

Community Attachment Indicators for 1995 to 1997

Two basic indicators of a household's community attachment were developed for the panel study. The first is the extent to which respondents reported that they attended family and village festivals and ceremonies. These festivals and ceremonies are a very old and traditional part of Russian rural life that survived, often in modified forms, during the Soviet era. Family ceremonies would include, for example, weddings, funerals and going away parties for young men who were going to serve in the military and when they returned home.

Village ceremonies and festivals would include May Day (May 1st), Veterans Day (May 9th), New Year's Day (January 1st), school graduation and harvest. Each village has a special local festival that is connected to the birthday of the patron saint of that village. Churches in all three villages were destroyed during Stalin's time. Residents of Latonovo currently are building a church. A relatively small number of villagers consider themselves members of a church, usually the Russian Orthodox Church; 8.3 percent in Latonovo, 21.2 percent in Vengerovka, and 20.0 percent in Bolshoe Sviattsovo. At present, in order to attend a church service residents of the villages have to travel to the nearest regional center, which involves travel by car or bus of between one-half and one hour each way. The percentage of respondents who had attended a religious service during the past year, reported in 1997, was 25.5, 37.8 and 22.0 in Latonovo, Vengerovka, and Bolshoe Sviattsovo, respectively.

A larger proportion of residents of the three villages read religious literature; 38.2 percent in Latonovo, 49.4 percent in Vengerovka, and 24.0 percent in Bolshoe Sviattsovo. The numbers of respondents who listen to religious programs on radio or television, by and large, broadcast by

Protestant evangelists, is even more substantial; 59.9 percent in Latonovo, 80.8 percent in Vengerovka and 62.7 percent in Bolshoe Sviattsovo. More than half of all respondents reported that they prayed during the past year; 51.6 percent in Latonovo, 60.3 percent in Vengerovka, and 50.0 percent in Bolshoe Sviattsovo.

There was a significant decline in overall attendance at family festivals and ceremonies during the three years of the panel study, as shown in Table 6.2. Those who reported that they attended these activities "often" declined from 24.2 percent in 1995 to 14.0 percent in 1997. This is consistent with ethnographic research in Orel *Oblast* that shows a general decline in the involvement of rural residents in the institutional life of their communities (Koznova 1996).

Residents in Latonovo and Vengerovka reported higher levels of participation in family festivals and ceremonies than residents of Bolshoe Sviattsovo during the entire three years of the panel study. These village differences reflect the greater social viability of villages in the *chernozem* zone of Russia today.

There was an even greater decline in attendance at village-level ceremonies and festivals. (See Table 6.3). The number of persons who reported that they never attended these types of festivals, rose from slightly more than one-third (35.9 percent) in 1995 to over one-half (51.4 percent) in 1997. The decrease in attendance at village-level festivals was greater in Latonovo and Vengerovka than in Bolshoe Sviattsovo. The lower level of decline in attendance at village-level festivals in Bolshoe Sviattsovo must be understood within the context of the much lower levels of attendance at these types of events in this village than in the other two villages at the beginning of the panel study.

98 Household Capital and the Agrarian Problem in Russia

Table 6.2: Attendance at Family Festivals and Ceremonies

Village	1995				1996				1997			
	Lat	Veng	Sviat	Total	Lat	Veng	Sviat	Total	Lat	Veng	Sviat	Total
N	157	156	150	463	157	156	150	463	157	156	150	463
Never	12.7	13.5	41.3	22.2	8.9	10.9	37.3	18.8	12.7	16.7	32.7	20.5
Sometimes	62.4	52.6	45.3	53.6	80.9	58.3	53.3	64.4	77.7	59.6	58.7	65.4
Often	24.8	34.0	13.3	24.2	10.2	30.8	9.3	16.8	9.6	23.7	8.7	14.0

Overall sample attendance at family festivals and ceremonies 1995-97: Kendall's Tau$_b$, p<.05; X^2 (4)=21.74, p<.001.
Village differences 1995:
X^2 (4)=54.09, p<.001 Village differences 1996: X^2 (4)=78.25, p<.001 Village differences 1997: X^2 (4)=37.76, p<.001

Table 6.3: Attendance at Village Festivals and Ceremonies

Village	1995				1996				1997			
	Lat	Veng	Sviat	Total	Lat	Veng	Sviat	Total	Lat	Veng	Sviat	Total
N	157	156	150	463	157	156	150	463	157	156	150	463
Never	37.6	17.9	52.7	35.9	56.7	25.0	57.3	46.2	51.6	37.2	66.0	51.4
Sometimes	56.1	48.1	40.7	48.4	40.1	46.8	39.3	42.1	46.5	53.8	31.3	44.1
Often	6.4	34.0	6.7	15.8	3.2	28.2	3.3	11.7	0.6	3.0	0.9	4.5

Total sample, attendance from 1995-97, Kendall's Tau$_b$, p < .001. Latonovo attendance from 1995-97, X^2(4)=14.90, p<.01.
Vengerovka attendance
from 1995-97, X^2(4)=34.48, p<.01.

Table 6.4 summarizes the overall decline in actual involvement in community institutional life during the three years of the panel study, as reflected in mean scores on the two-item community involvement index that combines responses to the two questions about participation in family and village festivals and ceremonies. This index will be used to measure community involvement in the structural equation models that are developed in subsequent chapters.

The decline in community involvement does not mean that this type of social capital is any less important as a correlate of household economic success in 1997 than it was in 1995. As we will see in Chapter 7, the index of community involvement continues to be associated with greater household production and sales throughout the course of the panel study. Nonetheless, *the overall decline in this type of social capital does raise some serious concerns about the relationship between traditional village attachments and the strategies that households are using to cope with social change.* This is a subject that will be discussed in later chapters.

Table 6.5 shows the mean scores for three years on the two-item index that measures the extent to which respondents report that they feel that they "fit" into their respective villages. This index, which will be used in the structural equation models of stress and mental health in Chapter 10 and quality of life in Chapter 11, measures a subjective sense of attachment to the local village community that does not necessarily correlate with actual levels of participation in village life. The two items contain scales, from 1 to 7, that indicate the extent to which a respondent feels that he or she can "relate" to others in the village and the extent to which he or she has much or little in "common" with other village residents. Responses to the two items were divided by two, so that the maximum possible score is seven.

Table 6.4: Mean Scores on the Index of Community Involvement (Range: 1-low to 6-high)

Village	1995				1996				1997			
	Lat	Veng	Sviat	Total	Lat	Veng	Sviat	Total	Lat	Veng	Sviat	Total
N	157	156	150	463	157	156	150	463	157	156	150	463
Mean Scores	3.81	4.37	3.26	3.82	3.48	4.23	3.18	3.63	3.47	3.79	3.13	3.47

Index of Community Involvement from 1995-97: $F(2)=11.43$, $p<.001$.

Community Attachment and Social Networks 101

Table 6.5: Mean Scores on the Index of Community Fit (Range 1-low to 7-high)

	1995	1996	1997
Latonovo n=157	5.05	5.07	4.99
Vengerovka n=156	4.85	4.77	4.88
Bolshoe Sviattsovo n=150	4.47	4.60	4.76
Total N=463	4.90	4.82	4.88

Differences between the villages in 1995: $F(2)=14.81$, $p<.001$; Scheffe comparison between Latonovo and Bolshoe Sviattsovo, $p<.01$; comparison between Vengerovka and Bolshoe Sviattsovo, $p<.01$. Difference in Community Fit in Bolshoe Sviattsovo from 1995 to 1997, $F(2)=25.32$, $p<.05$.

The mean score for sense of fit is close to 5 for the total sample during all three years. This mean is substantially higher than the scale average of 3.5. This shows that residents in the villages feel a high level of emotional attachment and belonging, in spite of the fact that their actual level of community participation has declined. Overall, there is no significant reduction in subjective level of attachment to the villages.

Two points, however, should be noted. First, in 1995 there was a significant difference between Latonovo and Vengerovka, on the one hand, and Bolshoe Sviattsovo on the other. By 1997, however, there was no significant difference between the three villages, which is due largely to the increase in the sense of fit in Bolshoe Sviattsovo. This is the only one of the three villages in which there is any statistically significant change in the mean level of subjective community fit over the three-year period of the panel study.

Second, the point cannot be overemphasized that even though actual levels of participation in these villages have declined in recent years, the village remains a very important source of satisfaction and a way of being "connected" for members of peasant households. This also parallels similar findings in American rural settings (O'Brien et al. 1994). Specific empirical evidence of the importance of this subjective connection on mental health and overall subjective quality of life will be shown in Chapters 10 and 11.

Social Network Indicators for 1995 to 1997

The mean sizes of the social exchange helping networks of households in the sample and the extent to which they changed during the panel study are shown in Table 6.6. In the total sample, all types of networks, with the exception of money exchange, trade and non-redundant ties increased from 1995 to 1997. This indicates that households made a substantial adjustment to a market economy, both in terms of instrumental and socio-emotional support. The reason that money exchange networks did not increase at the same rate as the other networks is simply because the high rate of inflation makes it extremely difficult for both borrowers and lenders to maintain a reasonably "fair" exchange rate. This is also true, to a lessor extent, with respect to trade of goods and services. There is an increase in trade networks in Vengerovka but not in the other two villages.

Alternatively, exchanges involving assistance with the plot or household chores are not subject to such inflationary pressures. The informal economy of peasant household production is largely outside of the formal economy and depends on the maintenance of informal exchange relationships. As stated in Chapter 1, we believe that social networks will have a substantial impact on household production and sales, and in turn on income. The effect of the changes in social networks, shown in Table 6.6, on household material well being, stress and subjective quality of life will be examined in Chapters 8 to 11. Thus, these types of social network arrangements have become an increasingly important part of the social capital of peasant households.

Finally, it should be noted that for the total sample it is the number of "redundant" rather than "nonredundant" ties that increased during the three years of the panel study. An increase in redundant ties means that the total number of persons in a household's social exchange network has not increased but the number of tasks or social support provided by the existing members of these networks has increased. Therefore, during this difficult transition, existing network members are being relied upon for more kinds of support.

At the start of the panel study in 1995, all types of networks, with the exception of trade and plot care, were higher in Latonovo than in Vengerovka and Bolshoe Sviattsovo. By 1997, however, there were no statistically significant differences between any of the networks in Latonovo and Vengerovka and differences between Latonovo and Bolshoe Sviattsovo in only three out of the eight social network exchange areas (household tasks, $p<.05$, discussion, $p<.01$ and non-redundant ties, $p<.05$). Table 6.6 shows that the major changes that occurred from 1995 to 1997 were increases in the size of networks in Vengerovka and Bolshoe Sviattsovo.

Table 6.6: Mean Social Network Indicators in 1995 and Changes from 1995 to 1997

	Latonovo (n=157)		Vengerovka (n=156)		Sviattsovo (n=150)		Total (n=463)	
	1995	95-97	1995	95-97	1995	95-97	1995	95-97
Money	1.84	-0.28	1.53	0.03	1.46	-0.08	1.61	-0.11
Trade	1.90	-0.20	1.62	0.42***	1.86	-0.03	1.79	0.07
Care	2.08	0.12	1.81	0.40***	1.46	0.51***	1.79	0.34***
Plot	2.41	0.22	2.15	0.68***	2.09	0.54***	2.22	0.48***
Household Tasks	2.02	0.21	1.45	0.53***	1.59	0.31*	1.69	0.35***
Discussion	2.36	-0.13	1.71	0.74***	1.74	0.04	1.94	0.25***
Total Redundant Ties	12.62	-0.06	10.25	2.81***	10.20	1.40*	11.04	1.38***
Total Non Redundant Ties	6.37	-0.31	5.28	0.69*	4.48	0.41	5.51	0.26

Levels of significance * p<.01, * p<.05, * p<.001.

The relationships between household labor, on the one hand, and the measures of the "purer" types of social capital on the other is shown in Table 6.7. In 1995, there were substantial differences in the amount of social capital possessed by households and the amount of household labor they possessed. During this time, household labor was positively associated with both types of community attachment. The relationship between household labor and community involvement was linear and the relationship between household labor and the subjective sense of fitting into the community was curvilinear. In the third year of the panel study, the relationship between household labor and community involvement remained strong. The relationship between household labor and sense of fitting into the community, however, had disappeared by this time.

The overall strength of association between household labor and the two measures of social networks in the total sample remained stable during the course of the panel study. The two biggest changes that occurred during the three years of the panel were the decline in community involvement and the increase in the number of redundant ties possessed by households in the sample. This is critical because it means that as time went on, more households shifted their energy toward building personal social networks of support that could help them to cope with a emerging market economy than broader community ties as indicated by less energy in participation in traditional festivals and ceremonies with other families and the whole village.

It is also important to observe that the most significant declines in community involvement occurred among those with the least and middle amounts of household labor. The biggest gains in developing redundant network ties were with the middle groups; those with household labor in the range of 1.75 to 2.74 and 2.75 to 3.74.

There are several implications in the shifts in household-level social capital that should be noted. First, during the Soviet period there was a great deal of institutional support from the *kolkhozy* for participation in village-level ceremonies and festivals. Not only did the collective farm provide material support for such celebrations, but the absence of opportunities for household economic growth meant that there was little competition for the energy and time of household members.

Table 6.7: Mean Community Attachment and Social Network Indicators by Level of Household Labor

	Community Involvement		Community Fit		Redundant Ties		Nonredundant Ties	
	1995	1997	1995	1997	1995	1997	1995	1997
0-1.74 (n=143)	3.36	3.03	4.63	4.88	10.13	10.85	4.75	4.90
1.75-2.74 (n=164)	3.76	3.56	4.81	4.88	10.82	12.59	5.40	5.86
2.75-3.74 (n=108)	4.28	3.80	5.00	4.88	11.78	13.65	6.14	6.51
3.75-4.74 (n=34)	4.32	3.80	4.74	4.84	12.35	13.77	6.47	6.30
4.75 + (n=14)	4.43	4.00	4.86	4.33	13.93	16.83	7.43	9.33
Total (N=463)	3.82	3.47	4.79	4.87	11.04	12.42	5.51	5.77

1995 - Household Labor and Community Involvement: $F(4)=12.48$, $p<.001$. 1995 Household Labor and Community Fit: $F(4)=2.36$, $p<.05$. Household Labor and Redundant Ties: $F(4)=6.07$, $p<.001$. Household Labor and Nonredundant Ties: $F(4)=9.48$, $p<.001$. 1997 – Household Labor and Community Involvement: $F(4)=12.18$, $p<.001$. Household Labor and Redundant Ties: $F(4)=9.17$, $p<.001$. Household Labor and Nonredundant Ties: $F(4)=11.45$, $p<.001$.

Second, economic opportunities for households have improved during the past few years. The absence of legal protections, especially the lack of government legal support for peasant household economic development has meant that households are more inclined to look inward rather than outward for protections. This translates into increased incentives to build highly personalized exchange networks and disincentives to participate in collective community-level festivals and ceremonies.

Thus, even during the short three-year period of the panel study, we witnessed a decline in the amount of social capital that was expended outside of the household (i.e., in community involvement), but an intensification of personal supportive network of the household. *The increase in redundant social network ties is especially noteworthy because it illustrates that households became increasingly more dependent on their own resources during periods of economic and legal uncertainty and were less willing to invest in outside community relationships.*

The aforementioned highlights a core dilemma facing rural areas of Russia with respect to the development of a civic culture, which is a key element in the long-term sustainability of democracy and market-type economic institutions. So long as legal and economic uncertainty continues, especially the former, household members, as rational economic actors, have little incentive to participate in institutional structures that are essential to the long-term development of democracy in Russia. This means that as the economic differentiation of households proceeds, according to their respective abilities to obtain household labor and purer forms of social capital, the resulting stratification system is not legitimized by citizen participation in the community-level institutions of civic culture. This is a crucial theme to which we will return later.

Summary

In this chapter, we examined three types of social capital, community involvement, sense of fit in the village and social networks, that have bearing on a Russian peasant household's economic and psychological adaptation to the emerging market economy. The vast majority of respondents in the panel study were born in the local area in which they currently live. Most of them have a substantial number of relatives living in their villages and they have lived in their villages for most of their lives. In this regard, the residents of these Russian villages are similar to rural residents in the American Midwest.

There is, nonetheless, variation between the three villages with respect to average length of residence. Residents in Bolshoe Sviattsovo, on average, have lived in their village less time than their counterparts in Latonovo and Vengerovka. This illustrates that community involvement of households in villages has become weaker in the *podzol* zone compared to households in the *chernozem* zone.

There are two main findings with respect to community involvement, as measured by participation in family and village ceremonies and festivals. The first, and most important, is that there has been a significant decline in community involvement from 1995 to 1997. The second is that there are significant differences between the villages, with lower levels of community involvement in Bolshoe Sviattsovo than in the other two villages. Again, this illustrates the higher social viability of the two black earth villages, Latonovo and Vengerovka. Nonetheless, as we will show in Chapter 8, the overall decline in levels of community involvement during the three years of the panel study has not diminished the importance of community involvement as a factor in facilitating a household's adaptation to a market economy.

Despite the decline in participation in village community life, most respondents continued to feel a part of their community life, as evidenced by the high scores on subjective sense of community fit. The major change that occurred in the three years was that the level of community fit increased in Bolshoe Sviattsovo to the point where village-level differences on this dimension were gone by 1997. In short, despite the stress generated by the need to adapt to the market economy, most households have remained emotionally connected to their villages. This is a point to which we will return in Chapters 10 and 11.

There was a substantial growth in the size of four out of six of the social exchange networks of village residents during the three years of the panel study, as well as an increase in the size of redundant tie networks. The latter means that although the number of persons in a respondent's networks did not increase from 1995 to 1997, those persons became involved in more exchange activities. This indicates an important adaptation of peasant households to an emerging market economy.

The significance of the growth in peasant household social networks has to be evaluated in light of the decline in community involvement. The findings show that, overall, peasant households have elected to devote more energy to building personalized and multi-plex social networks that are directly connected to household tasks, while, at the same time, devoting less energy to participation in more comprehensive village-level activities. The implications of these trends for the development of democratic institutions in rural Russia will be discussed in Chapter 12.

Finally, there are strong positive associations between levels of household labor and the ability of a household to obtain various types of social capital. Community involvement remains strongly associated with levels of household labor throughout the panel study. The subjective sense of community fit is associated with household labor in 1995, but this association disappears by 1997. The association between both measures of overall social network strength, nonredundant and redundant ties, and household labor remains from 1995 to 1997.

7 Physical Capital

David J. O'Brien, Valeri V. Patsiorkovski and Larry D. Dershem

Introduction

In this chapter, we will examine three types of physical capital that have an impact on the ability of peasant households to produce and sell agricultural products; land, tools and livestock. There are three major sets of constraints that affect an individual peasant household's access to any of these types of physical capital, each of which is associated with a different level of analysis.

At the macro-level, the historical relationships between peasant households and *kolkhozy* and *sovkhozy* have provided limited opportunities for households to obtain any of these types of capital. As shown in Figure 1.1 in Chapter 1, even in 1996 large enterprises controlled most agricultural land in Russia. Household plots accounted for 46 percent of all domestic agricultural production, but they cultivated a mere 4.2 percent of all arable land. Similarly, large enterprises have had a virtual monopoly on control of expensive agricultural equipment, such as tractors and combines. Large enterprises have had less of a monopoly on livestock, largely because the survival of many of these enterprises depends on their focusing on large-scale grain production. Yet, some large enterprises own substantial herds of beef and dairy cattle and some large enterprises have large numbers of fowl and hogs for factory-like production operations.

Nonetheless, macro-level changes in Russian agriculture and in Russian society as a whole since 1991 have provided a small window of opportunity for households to increase their physical capital and thereby to expand their production potential. Despite the impasse between the executive branch and Duma in the Russian central government over land reform, some opportunities for peasant households to gain access to small quantities of land through rental arrangements have become a reality in the past few years. In addition, as restrictions on household production and sales have loosened, households now have the resources to purchase or trade for some types of agricultural equipment. Finally, as the large enterprises have become less able to pay cash for work by peasant household members, increasingly they have paid their workers in grain. This increase in grain has permitted peasant households to substantially increase the amount of animals they own.

At the mezzo- or village-level, however, there is considerable discretion with respect to the ways in which opportunities for improving peasant household production may be implemented. The main opportunity for land rental, renting land ceded to the local government by the *kolkhoz* or *TOO*, has been implemented differently in individual villages, thereby creating different opportunities for expansion of household production from one village to another. There also are differences in ownership of agricultural tools and animals in different villages.

The most striking differences in access to physical capital, however, have been at the micro- household-level. Some households have been able to translate advantages in human and social capital into increased access to the various forms of physical capital. Differential access to human and social capital creates differences in the ability of a household to earn enough money to rent land, to buy agricultural tools, and to purchase and care for animals. Moreover, as we described in Chapter 2, these relationships between human and social capital, on the one hand, and physical capital, on the other, are critical in reinforcing initial advantages or disadvantages that a household possesses as its members try to compete in an emerging market economy. These differences, as well as the different opportunities for access to physical capital provided by individual villages play an important role in the system of economic inequality that is emerging in Russian rural villages.

Land Tenure and Use

In the Soviet period, land use was restricted to the large tracts that were controlled by the *kolkhozy* or *sovkhozy* and the small garden plots that were owned by peasant households. Technically, the *kolkhozy* and *sovkhozy* differed in that the *kolkhozniky* (collective farm workers) "owned" the land in the former, while the land in the latter was owned by the state. In practice, land use in both the *kolkhozy* and *sovkhozy* was decided by their

Physical Capital 111

management, in response to centralized plans generated originally in Moscow by the government, which were then implemented down the line into the various regions and *oblasts*.

Throughout the Soviet period, peasant households were permitted to own a small plot of land surrounding their dwellings, usually not more than one-third hectare in size, on which they raised crops and animals that could be sold in local farmers' markets. As noted earlier, these plots produced a considerable portion of total domestic agricultural output during the Soviet period. It is important to recognize that throughout the Soviet period, peasant household plot production was dependent on positive relationships between the household and the *kolkhoz* or *sovkhoz*. Peasant households, for example, depended on tractor drivers from the large enterprises to plow their fields in the spring. In addition, the households received a substantial proportion of their inputs, such as seeds and fertilizers from the large enterprises (O'Brien et al. 1998b: 134-137).

The average size of the land used by peasant households for production in the three villages in the study increased after 1993, but, by Western standards, remained quite small. The average total size of land used by households in Latonovo and Bolshoe Sviattsovo was slightly over one hectare in 1995 and the corresponding figure for Vengerovka was one-half hectare at that time. The number of households using more than one hectare of land increased in all three of the villages from 1993 to 1995, but decreased from 1995 to 1996. At the time the panel study began in 1995, less than two percent of the households in the three villages were cultivating ten hectares or more. A total of fourteen households in the three villages were officially registered as private farmers (*fermery*). The average size of their land holdings was 21.37 hectares, which is considerably larger than the average peasant household but still quite small by West European or American farming standards (O'Brien et al. 1998b: 48-49).

Table 7.1 shows the distribution of the three main types of land use in the three villages during the three years of the panel study. Household plots, the traditional garden plots that peasant households were permitted to use and sell products from during Soviet times, are, on average, slightly less than one-third of a hectare in size. The plots in Vengerovka and Bolshoe Sviattsovo are approximately twice as large as those in Latonovo. The larger size plots in Bolshoe Sviattsovo does not present any real advantage to peasant households in that village, compared to households in Latonovo since Bolshoe Sviattsovo lies outside of *chernozem* zone and thus has poorer soil and a shorter growing season. Both Latonovo and Vengerovka, however, are located in the *chernozem* zone, with similar soil and climate conditions. Thus, the larger size of the household plots in Vengerovka does produce a competitive advantage for households in that village compared to households in Latonovo.

Table 7.1 also shows the amounts of two new types of land tenure that are found in post-Soviet Russian villages. The first is rental of land. This can take two forms, one formal-legal and the other informal. Formal rental of land was not permitted until 1991 when President Yeltsin issued a decree stating that each large enterprise must cede a small portion of its land to the local government of the village in which it operated. The local government in each village then had discretion about how such land would be distributed to peasant households. In the villages of Vengerovka and Bolshoe Sviattsovo, peasant households were permitted to rent small pieces of land from the local government. President Yeltsin has issued successive decrees in subsequent years to permit this arrangement to continue. Most rented land is based on this type of formal agreement.

Informal rental of land, although not measured in this study, which is not sanctioned by law, involves *informal* agreements with neighbors in which a household agrees to return a certain portion of the harvest to neighbors in return for use of their land. This type of arrangement is most common in the villages where there are no family members to care for household plots owned by elderly widows or in cases where there are many temporary or seasonal residents. This type of arrangement is more common in Bolshoe Sviattsovo than in the other two villages. This is due to the fact that Bolshoe Sviattsovo does not have as strong family ties as do the other two villages and Bolshoe Sviattsovo also has a larger number of temporary and seasonal residents.

Both forms of rental arrangements currently used in the villages are not the direct result of the new land reform and privatization scheme, which are geared toward reforming large enterprises and not households. Rather, most of these rental arrangements are the direct result of households' strategies, in cooperation with local governments, to survive in a changing economic environment.

Table 7.1: Mean Size of Different Types of Land Use by Village and by Year

	Household Plot			Rented Land			Communal Land		
	1995	1996	1997	1995	1996	1997	1995	1996	1997
Latonovo (n=157)	0.17	0.15	0.15	0.00	0.00	0.00	0.69	0.70	0.70
Vengerovka (n=156)	0.38	0.39	0.40	0.13	0.13	0.16	0.00	0.00	0.00
Sviattsovo (n=150)	0.40	0.42	0.42	0.12	0.11	0.20	0.00	0.00	0.00
Total Sample (N=463)	0.32	0.32	0.32	0.08	0.08	0.12	N/A	N/A	N/A

Total Rental Land from 1995-97: $F(2)=5.09$, $p<.01$. Rental Land by Village, 1995: $F(2)=31.37$, $p<.001$. Rental Land by Village, 1996: $F(2)=21.94$ $p<.001$. Rental Land by Village, 1997: $F(2)=33.18$, $p<.001$.

According to Russian law, through the privatization and land reform program implemented from 1992 to 1994, peasant households are the formal-legal owners of a large portion of arable land in rural areas. In practice, most peasant households, for various reasons, do not have physical access to this land for personal use. Russian law (Decrees No. 1761 in 1993 and No. 337 in 1996) provides peasant households with several ways to receive benefits from this land: 1) registering as *fermery* and using this land as part of their business, 2) using it for household production as they use their household plots, 3) rent this land to other agricultural enterprises [such as a *kolkhoz*, *TOO*, *fermery*], (see Appendix 11 for a copy of the land rental agreement), 4) sale, 5) placing it "under trust management" (*doveritelnoe upravlenie*), or 6) placing this land in the local enterprise's, *kolkhoz or TOO*, charter fund (*ustavnoii fond*). However, if this last option is chosen, ownership of this land is transferred from the household to the enterprise.

In our three villages, in 1997, approximately 90 percent of the households rent this land to the large enterprises (*kolkhoz* or *TOO*), and 10 percent of households rent this land to *fermery* or use this land for household production. Similar findings are reported in a study of eleven *oblasts*. In this study of 364 large enterprises, 90.7 percent of landowners chose to rent this land back to the large enterprise. Only 1.5 percent transferred ownership to the local enterprise's charter fund and only 1 percent sold it. The remaining 5 percent chose other options or simply refused to accept ownership of land (*Programma privatizatsii i reorganiizatsii sel'skokhoziaistvennykh predpriiatii* 1998).

The second type of new land tenure arrangement, in our study, is communal land. This arrangement is based on the same Presidential decrees that ceded land from the large enterprises to local village governments. In some villages, the local government decided that some or the entire portion of the land that it controlled would be set-aside as "communal land." Each household would be responsible for its portion of the cultivation of the land and each household would receive a share of the harvest from that land. Decisions about what to plant and how harvest shares would be allocated were made collectively. Thus, this communal land has become, in practice, a "mini-*kolkhoz*."

This type of land tenure arrangement is found only in one village, Latonovo. In Latonovo, the share each household was allotted in the communal production plot was decided by a formula in which it could only

Physical Capital 115

have total "ownership" of 0.9 hectare of land *that included both the household plot and their share of the communal land.* Thus, if a household had a private plot that was 0.40 hectare in size, its share of the communal land was 0.50 hectare. If the household's private plot was 0.20 hectare in size, its share of the communal land was 0.70. In short, there was no incentive for a household to increase its productive capacity. This, as we will see in the next chapter, created disincentives for peasant households in Latonovo to increase their household production and sales.

The most important type of land tenure arrangement to emerge in the post-Soviet Russian countryside is the rental arrangement. Most of the rented land referred to in Table 7.1 is based on legal contracts between households and local governments in Vengerovka and Bolshoe Sviattsovo. Rental of land has become a very important way in which peasant households can increase their productive capacity in spite of the failure of the Russian Central Government to resolve the land ownership issue (O'Brien et al. 1998b; Wegren 1998). Although the mean size of rented land is still relatively small, it nonetheless has grown during the short three-year period of the panel study.

Many critics of current privatization and land reform efforts in Russia view the changes we have described as merely rhetorical and unimportant. Nevertheless, we believe that these small efforts will have substantial consequences in changing land relations among households and other enterprises in rural areas. The establishment of rental contracts between households and the local Village Councils is an illustration of the evolution of formal constraints (North 1990:46-53) and institutional protections that encourage economic development.

Rental of land historically has been an important factor in the development of small farms in other national settings. This has been true in situations where legal restrictions have prevented certain groups from owning land. This was the case with Japanese Americans on the West Coast during the pre-World War II period. Although they were prevented by "alien land laws" from legally owning land, they nonetheless developed a very successful *labor intensive* agricultural system in the Central Valley of California, relying on informal rental arrangements with Caucasian neighbors (O'Brien and Fugita 1991:19-37).

More recently, in the American Midwest, the high cost of land and fear of debt, which bankrupted many middle-sized farms during the "farm crisis" of the mid-nineteen eighties, has encouraged many American Mid-

Western farm families to cultivate a substantial portion of rented land for their production. In 1992, for example, 44 percent of all cropland in the United States was rented. Since the 1970s, rented land, as a percentage of total farm acres, has increased, on average, about 3.5 percent every five years (U. S. Department of Commerce 1984; U. S. Department of Commerce 1994).

There is a significant increase in the amount of land rented by peasant households during the three-year time period of the panel study. Rental arrangements involving peasant households, however, varied considerably from one village to another. This type of land use arrangement was completely absent in Latonovo during the three years of the panel study, but increased in Vengerovka and Bolshoe Sviattsovo during the same time period. The slightly larger amount of rented land in Sviattsovo than in Vengerovka does not present any advantage to the former over the latter since Sviattsovo's soil and climate is much less productive than Vengerovka's. But, the larger amount of rented land in Vengerovka, compared to Latonovo, does give the former a substantial advantage over the latter in terms of productive potential. This condition, coupled with the easier access of peasant households to credit in Vengerovka, compared to Latonovo, creates structural advantages for households in Vengerovka that will become evident when we examine household production and sales in the next chapter.

Table 7.2: Mean Size of Rented Land by Household Labor

Weighted No. of Adults	1995	1996	1997
0-1.74	0.0 (n=143)	0.0 (n=146)	0.0 (n=147)
1.75-2.74	0.0 (n=164)	0.0 (n=177)	0.15 (n=173)
2.75-3.74	0.14 (n=108)	0.12 (n=95)	0.14 (n=93)
3.75-4.74	0.0 (n=34)	0.13 (n=35)	0.23 (n=44)
4.75+	0.12 (n=14)	0.0 (n=10)	0.0 (n=6)
Total Sample (N=463)	0.00	0.00	0.12

Rented land by weighted number of adults: 1995-$F(4)=8.17$, $p<.001$; 1996-$F(4)=4.84$, $p<.001$; 1997-$F(4)=7.96$, $p<.001$.

The amount of land rented is associated with the amount of labor available in a household as shown in Table 7.2. The number of cases within each category of household labor are different for each year because some

Physical Capital 117

households in the panel sample have been reconstituted, gaining or losing members, and thus have changed the amount of household labor they possess. This relationship remains throughout the three years of the panel study. None of those with the least amount of household labor, scores of 0 to 1.74, report renting any land. These are mainly widows and retired couples. Over the three-year period, the households that are most consistent in renting land are those with household labor scores of 2.75-3.74 and 3.75-4.74. Those households with the highest level of labor, scores of 4.75 and above, reported renting land during 1995, but do not report any renting of land during the last two years of the panel study. Additional analysis shows that it is the fourth and fifth demographic types of households, employed couples with children and employed couples with children and other adults, that comprise the two household labor groups that are most likely to rent land.

Table 7.3: Mean Percentage of Rented Land by Level of Community Involvement

Community Involvement	1995	1996	1997
1.00-2.99 (low)	0.00 (n=83)	0.0 0(n=75)	0.0 0(n=88)
3.00-5.99 (middle)	0.0 0(n=326)	0.0 0(n=346)	0.12 (n=355)
6.00-7.00 (high)	0.20 (n=54)	0.20 (n=42)	0.18 (n=20)
Total Sample (N=463)	0.00	0.00	0.12

Rent by Community Involvement: 1995-$F(4)=14.04$; $p<.001$; 1996-$F(4)=11.14$, $p<.001$; 1997-$F(4)=2.83$, $p<.06$.

The relationship between levels of community involvement and a household's propensity to rent land is presented in Table . Although overall levels of community involvement declined substantially during the three years of the panel study (see Chapter 5), the positive relationship between community involvement and household rental of land remained from 1995 to 1997, but the strength of that relationship declined over the three- year period. In 1997, those households with a middle level of involvement in village- and family- ceremonies and festivals were more likely to rent land in that year than they were during the preceding two years. The most important substantive point is that those households that are not involved at all in the ceremonies and festivals of their neighbors or the community-as-a-whole do not rent land. Alternatively, households with the highest level of community involvement had the highest levels of land rental during the three years.

Table 7.4: Rental of Land by Number of Non-Redundant Ties from 1995 to 1997

Number of Non-Redundant Ties	1995	1996	1997
0-3	0.00 (n=102)	0.00 (n=91)	0.00 (n=89)
4-5	0.00 (n=155)	0.00 (n=145)	0.00 (n=133)
6	0.15 (n=66)	0.10 (n=74)	0.19 (n=79)
7-10	0.00 (n=122)	0.00 (n=139)	0.15 (n=144)
11 – 17	0.00 (n=18)	0.00 (n=18)	0.00 (n=18)
Total Sample (N=463)	0.00	0.00	0.12

Rent by nonredundant ties: 1995-$F(4)= 3.89$, $p<.01$; Scheffe, 0-3 x 6, $p<.01$; 1996- $F(4)=1.22$, p=n.s.; 1997 $F(4)=6.05$, $p<.001$, Scheffe, 0-3 x 6, $p<.01$; 0-3 x 7-10, $p<.05$, 4-5 x 6, $p<.05$, 0-3 x 7-10, $p<.05$.

Table 7.4 shows that there is a curvilinear relationship between the size of a household's non-redundant networks and whether or not it rented land in 1995 (similar findings for redundant ties are shown in Appendix 7). In that year, middle-sized networks (6 to 10 members) were more likely than either small or very large networks to rent land. It comes as no surprise that small networks would be less likely to facilitate a household's ability to obtain and work more land. What is perhaps surprising, however, is that households reporting the highest number of non-redundant ties also were unlikely to rent land in 1995. The explanation for the latter is simply that in 1995 many of the households that were made up of elderly couples had a large number of persons in their helping networks. These helping networks, however, are oriented largely to helping these individuals overcome the limitations of age and not to increasing their amount of rented land.

Alternatively, households in the middle-range of helping networks are more likely to contain younger couples and their children and other relatives (demographic types four and five). These households have a smaller number of persons in their networks but the assistance of these persons is more likely to be associated with increasing the household's productive capacity. This point will be illustrated in the next chapter when we examine the relationship between network size and overall household production and sales.

By 1997 households with the largest number of non-redundant ties began to rent more land than they did in 1995 or 1996, although the mean level of land rented by these types of households still is lower than households with middle-sized helping networks.

Agricultural Tools

The percentage of households owning various types of agricultural equipment and tools from 1995 to 1997 is shown in Table 7.5. The frequency of ownership of tractors is quite small, never comprising more than three and one-half percent of the households in the sample. Similarly, only a small number of households own motorized agricultural equipment.

Table 7.5: Percentage of Households Owning Agricultural Equipment and Tools in 1995, 1996 and 1997

Agricultural Equipment	1995	1996	1997
Tractors	3.2	3.2	3.4
Agricultural Motors	1.3	2.0	2.3
Agricultural Tools	9.5	8.9	12.3
Automobiles	16.4	21.6	21.8
Motorcycles	24.0	25.5	26.3

Most peasant households rely upon the *TOO* or *kolkhoz* to provide cultivation of their plots during the spring. Typically, households take down their fences and the *kolkhoz* tractor driver goes through a row of household garden plots. Each household pays a certain amount to the tractor driver for these services. The most important source of variation between peasant households is whether they possess agricultural tools that facilitate production or processing. This includes, for example, fruit sprayers and cream separators. The ability to make cream is especially important because it is the first step in the process of producing sour cream or butter, value-added products that can yield considerable income for a peasant household. Slightly more than one out of ten households (12.4 percent) have at least one piece of equipment of this type.

If we only consider those households that have at least some capability for substantial agricultural production, however, the proportion of households having some agricultural equipment is substantial. If we do not include single person (mainly elderly widows) and retired couple households in our calculations, for example, then the proportion of households having some type of agricultural equipment is 18.3 percent, or close to one-fifth of the potentially productive households in the villages.

Also included in Table 5 is the frequency of ownership of automobiles and motorcycles by households during the three years of the panel study. Automobiles and motorcycles are not convenience items, nor are they used for commuting to work. Rather, in the Russian peasant household economy these vehicles are used to transport hay and grain from fields, as well as animals to market. Motorcycles typically have sidecars and automobiles typically pull trailers. Thus, if a household can obtain either an automobile or a motorcycle they have substantially increased their productive capacity. In this regard, it is noteworthy how much ownership of automobiles increased from 1995 to 1997, from 16.4 percent to 21.8 percent.

Although the social organization and dependence on hand labor in the Russian peasant household is similar to peasant households elsewhere in the world, the level of education and technological skills is much higher in the Russian villages. In order to maintain automobiles or motorcycles in villages with very limited facilities for repair, Russian peasant households need to have a substantial store of technical knowledge and to be able to learn new procedures and techniques for repairing increasingly complex engines.

Table 7.6: Percentage of Households Using Selected Types of Agricultural Inputs in 1995, 1996 and 1997

Agricultural Input	1995	1996	1997
Hybrid seed	6.3	4.1	2.6
Hybrid animals	1.7	2.4	1.5
Chemical Fertilizers	3.5	1.9	4.3
Organic Fertilizer	87.3	86.6	85.5
Herbicides/pesticides	49.9	60.3	58.5
Greenhouse	49.9	60.3	58.5

Table 7.6 shows the percentage of households in the villages that use various kinds of agricultural inputs. The vast majority of households use organic rather than chemical fertilizers, which are obtained from their herds and flocks. The main types of advanced agricultural technology that are

used by peasant households are herbicides and pesticides, mainly chemicals to control diseases that affect potatoes. The proportion of households that use hybrid seeds or advanced breeding practices for animals remains very small. Even with low quality inputs, however, these households have managed to achieve a remarkably high level of production (Nazarenko 1998:10).

Peasant household production in these Russian villages typifies the labor-intensive practices used in peasant household production in other parts of the world. There is, however, one important caveat. All of these households do depend upon the *TOO* or the *kolkhoz* to assist them in cultivation, as well as with other phases of their household production process.

Table 7.7: Mean Number of Total Agricultural Tools (excluding tractors) by Village and Year

Village	1995	1996	1997
Latonovo (n=157)	0.15	0.24	0.30
Vengerovka (n=156)	0.00	0.00	0.12
Sviattsovo (n=150)	0.22	0.27	0.39
Total Sample (N=463)	0.14	0.19	0.27

Village Differences in 1995: $F(2)=5.14$, $p<.01$. Village Differences in 1996: $F(2)=4.69$, $p<.01$; Village Differences in 1997-$F(2)=5.74$, $p<.01$.

Table 7.7 shows the extent to which households have agricultural tools in each of the three study villages. Households in Latonovo and Bolshoe Sviattsovo, on average, have more agricultural tools than households in Vengerovka. Moreover, while households in Latonovo and Bolshoe Sviattsovo increased their ownership of tools during each of the three years of the study, it was not until 1997 that households in Vengerovka obtained any appreciable number of agricultural tools. This is especially important to note since, as we saw earlier, some households in Vengerovka rent additional land for household production while this situation does not exist in Latonovo. This means, therefore, that, on average, households in Vengerovka are working more land by themselves without the assistance of equipment.

122 Household Capital and the Agrarian Problem in Russia

Table 7.8: Mean Number of Total Agricultural Tools (excluding tractors) by Household Labor and Year

Weighted No. of Adults	1995	1996	1997
0-1.74	0.00 (n=143)	0.00 (n=146)	0.00 (n=147)
1.75-2.74	0.20 (n=164)	0.26 (n=177)	0.34 (n=173)
2.75-3.74	0.21 (n=108)	0.27 (n=95)	0.41 (n=93)
3.75-4.74	0.12 (n=34)	0.29 (n=35)	0.39 (n=44)
4.75 +	0.14 (n=14)	0.30 (n=10)	0.50 (n=6)
Total Sample (N=463)	0.14	0.19	0.27

Tools by Household Labor in 1995: $F(4)=3.34$, $p<.01$. Tools by Household Labor in 1996: $F(4)=3.52$, $p<.01$; Tools by Household Labor in 1997: $F(4)=5.23$, $p<.001$.

Table 7.8 shows that there is a positive relationship between household labor and the acquisition of agricultural tools during all three years of the panel study. Not only are households with higher levels of household labor more apt to have agricultural tools, but the advantages of having more household labor, especially comparing the lowest category (0-1.74) with the remaining four categories, become more evident during each successive year of the panel study.

The gain in the mean number of total agricultural tools becomes substantially greater as the amount of household labor increases. The mean level of agricultural tools for households in the second category of household labor (1.75-2.74) increased less than the amount reported in 1995, but the mean level of agricultural tools reported by households in the third category (2.75-3.74) more than doubled from 1995 to 1997. Households in the fourth (3.75-4.74) and fifth (4.75 +) categories experienced even greater gains of more than three times the level of agricultural tools they reported possessing in 1995.

Table 7.9: Mean Number of Total Agricultural Tools (excluding tractors) by Level of Community Involvement by Year

Community Involvement	1995	1996	1997
1.00-2.99 (low)	0.00 (n=83)	0.00 (n=75)	0.10 (n=88)
3.00-5.99 (middle)	0.14 (n=326)	0.23 (n=346)	0.31 (n=355)
6.00-7.00 (hi)	0.28 (n=54)	0.24 (n=42)	0.25 (n=20)
Total Sample (N=463)	0.14	0.19	0.27

Tools by Community Involvement: 1995- $F_{(2)}=3.19$, $p<.05$; 1996- $F_{(2)}=3.19$, $p<.05$; 1997- $F_{(2)}=2.85$, $p<.06$.

Table 7.9 shows that there is a positive relationship between community involvement and a household's acquisition of agricultural tools. However, this relationship operates indirectly through the association between community involvement and rental of land (see Table 7.3).

The mean number of total agricultural tools for households with different levels of redundant and nonredundant social exchange network ties is shown in Appendixes 8 and 9. The only statistically significant relationship is between redundant ties and tools in 1997. This relationship is curvilinear. The propensity to own tools increases with the number of redundant ties but begins to decline in households with the highest levels of redundant ties. This pattern is similar to the one observed earlier, where the strongest relationship was found between the middle level of ties and rental of land.

Livestock

As noted in Chapter 1, ownership of livestock is expected to be a major source of differentiation between peasant households in Russian villages. Meat and dairy products, especially value-added products, are the major source of cash and barter for these households. Thus, the ability of a household to maintain herds and flocks of animals will be an important predictor of the material quality of life of its members.

124 Household Capital and the Agrarian Problem in Russia

Table 7.10: Percentage of Households Owning Different Amounts of Selected Types of Livestock (N=463)

Number Owned	Cows			Pigs			Poultry		
	1995	1996	1997	1995	1996	1997	1995	1996	1997
0	38.2	39.5	41.7	36.1	33.7	36.9	11.4	8.9	11.9
1-2	52.3	51.6	45.8	49.3	44.7	42.1	0.9	0.6	0.2
3-4	8.9	8.0	10.0	11.9	15.3	12.3	2.8	2.2	1.5
>4	0.6	3.8	2.5	2.8	6.3	8.7	84.9	88.3	86.4

Increases in Livestock from 1995-97. Pigs: $F(2)=4.66$, $p<.010$. Poultry: $F(2)=4.28$, $p<.01$.

Table 7.10 shows the proportion of households with different numbers of cows, pigs, and poultry from 1995 to 1997. There is no significant increase in the numbers of cows owned by villagers during this period, but there are significant increases in household ownership of pigs and poultry. Pigs and poultry are especially important in the Russian peasant household economy because they can be maintained largely with hand labor and do not require the amount of space that would be necessary to keep larger animals, such as cattle. Thus, a household can maintain pigs and poultry in relatively large numbers, restricted only by its capacity to obtain grain or potatoes to feed these animals and its capacity to provide hand-labor to care for them. Moreover, pigs and poultry offer an opportunity for a very quick turn-around in profit, from birth to slaughter, which means that the household does not have to tie up its very limited financial capital for very long.

Table 7.11: Mean Level of Household Ownership of Livestock in the Three Villages

	Cow			Pig			Poultry		
	1995	1996	1997	1995	1996	1997	1995	1996	1997
Latonovo (n=157)	0.9	0.9	0.9	1.5	2.0	1.8	18.5	20.4	24.0
Vengerovka (n=156)	1.1	1.1	1.2	1.7	2.0	2.3	40.5	44.4	52.1
Sviattsovo (n=150)	1.3	1.3	1.5	0.7	1.0	0.8	9.1	10.3	10.1
Total Sample Mean (N=463)	1.1	1.1	1.2	1.3	1.7	1.6	22.9	25.2	29.0

Cows by Village: 1995- $F_{(2)}=5.18$, $p<.01$; 1996- $F_{(2)}=3.05$, $p<.05$; 1997- $F_{(2)}=5.55$, $p<.01$. Pigs by Village: 1995- $F_{(2)}=23.59$, $p<.001$; 1996-$F_{(2)}=9.17$, $p<.001$; 1997-$F_{(2)}=22.60$, $p<.001$. Poultry by Village: 1995-$F_{(2)}=110.05$, $p<.001$; 1996-$F_{(2)}=99.17$; 1997-$F_{(2)}=130.84$, $p<.001$.

Table 7.11 shows that household ownership of livestock varies considerably from one village to another. Households in Bolshoe Sviattsovo, for example, had more cows than Latonovo or Vengerovka during the three years of the panel study. This reflects the specific type of land and climate conditions in Bolshoe Sviattsovo that favor dairy operations.

The number of pigs and poultry, however, are substantially higher in Vengerovka than in the other two villages. In Vengerovka, peasant households typically have large flocks of chickens, ducks and geese that are confined behind fences in the household during the night but leave the premises during the day. Ducks and geese are led to a common pond area during the day and return to the household at night. In addition, households in Vengerovka typically keep at least three pigs until they need money for a significant purchase, such fuel for the winter or clothing for the children to attend school in the fall. Both the poultry and pigs are fed by grain that the household receives as part of its share for working on the *TOO*.

Table 7.12: Mean Weighted Number of Animals in Households by Village

Village	1995	1996	1997
Latonovo (n=157)	323.93	351.32	344.97
Vengerovka (n=156)	396.94	412.34	474.84
Sviattsovo (n=150)	378.07	373.31	414.05
Total Sample (N=463)	366.07	379.00	411.11

Weighted number of animals by village in 1997: $F(2)=4.12$, $p<.05$.

Table 7.12 shows the weighted number of animals according to the procedures described in Chapter 4. The figures shown are summaries that give a weight to each animal by its relative market value; 1 for poultry, 50 for a pig and 250 a cow. For the total sample there is no overall change in number of animals owned by households during the three years of the panel study. In addition, the only year in which there is a statistically significant difference between villages is 1997, and this difference, due to the larger number of animals in Vengerovka than Latonovo, is not large. There are, however, substantial differences between households within each village that are based on differences in the amount of human and social capital that they possess.

Table 7.13: Mean Weighted Number of Animals by Household Labor

Weighted No. of Adults	1995	1996	1997
0-1.74	102.22 (n=143)	105.92 (n=146)	116.54 (n=147)
1.75-2.74	435.84 (n=164)	462.72 (n=177)	498.27 (n=173)
2.75-3.74	503.25 (n=108)	526.01 (n=95)	546.33 (n=93)
3.75-4.74	618.91 (n=34)	653.17 (n=35)	746.50 (n=44)
4.75+	571.43 (n=14)	528.00 (n=10)	559.00 (n=6)
Total Sample (N=463)	366.07	379.00	411.11

Weighted Number of Animals by Household Labor: 1995-$F(4)=55.37$, $p<.001$; 1996-$F(4)=47.83$, $<.001$; 1997-$F(4)=44.06$, $p<.001$.

Table 7.13 shows a very strong relationship between the amount of labor in a household and its ability to own and care for livestock. Care of livestock is the most labor-intensive part of peasant household production. Households with more labor obviously have an advantage over other households. The greatest gain in numbers of animals is from the first to the second category of household labor. Households with scores of 1.75-2.74 have more than four times the number of animals that are owned by households with scores from 0-1.74. There are consistent gains from the second to the third categories, 15 percent, and from the third to the fourth category, 23 percent. There is a slight decline from the fourth to the fifth category, 8 percent, but the overall trend is for a household's capacity to own and care for animals to increase as its labor potential increases.

Care of animals requires a great deal of cooperation between the peasant household and other members of the village community. A significant portion of the livestock owned by households uses common land, including pastures for cows and ponds for ducks and geese. In addition, the household typically needs help from neighbors to watch livestock when they have to go out of town, for business or family matters, or when members of the household are sick. Moreover, other persons in the village, as well as the leadership in the *TOO* or *kolkhoz* are more likely to be called upon for assistance with inputs, cultivation, or marketing when the household has more livestock. In short, having livestock makes the household more dependent on the community as a whole and thus provides an incentive to maintain family involvement in community festivals and ceremonies.

Table 7.14: Mean Weighted Number of Animals by Community Involvement

Community Involvement	1995	1996	1997
1.00-2.99 (low)	277.77 (n=83)	216.52 (n=75)	194.15 (n=88)
3.00-5.99 (middle)	350.88 (n=326)	391.13 (n=346)	449.27 (n=355)
6.00-7.00 (high)	593.50 (n=54)	569.24 (n=42)	688.35 (n=20)
Total Sample (N=463)	366.07	379.00	411.11

Weighted Number of Animals by Community Involvement: 1995-$F(2)=18.30$, $p<.001$; 1996-$F(2)=15.05$, $p<.001$; 1997-$F(2)=20.70$, $p<.001$.

Table 7.14 shows that the extent to which a household is involved in the village community, is strongly associated with its capacity to own and care for animals. Moreover, this relationship becomes stronger over time. By 1997, households that are highly involved in the village community, as measured by their attendance at family and village festivals and ceremonies, have, on average, three and one-half times more animals than households with the lowest level of involvement in village community life.

Appendixes 9 and 10 show that households with more redundant ties had more animals in 1995 and 1997. A positive relationship between nonredundant ties and weighted number of animals was found in all three years of the panel study. These relationships, however, disappear when the other social capital variables are included in the structural equation models that are shown in the next chapter.

Summary

In this chapter we examined the distribution of the main types of physical capital that affect the ability of a peasant household to produce and sell agricultural commodities. These types of physical capital are conceptualized as intervening variables, between the household's human and social capital, on the one hand, and its ability to compete in a new market economy, on the other. In their role as intervening or intermediary mechanisms in the overall production process, these forms of physical capital provide additional advantages to households that already have advantages in human and social capital. Thus, physical capital can further increase the differentiation of households in Russian villages.

Despite the impasse over land reform in the central government, the data show a differentiation in rental of land by village and by household. Rental of land by households is found in Vengerovka and Bolshoe Sviattsovo but not in Latonovo. This is a very important illustration of how local choices do exist and that these decisions have practical implications for the quality of life of peasant households.

Overall, households with more household labor, higher levels of community involvement and more nonredundant network ties were more likely to rent land than were households with lower levels of labor, community involvement and helping networks.

The relationship between community involvement and rental of land is linear. Despite the overall reduction in levels of community involvement in the Russian villages, households with higher levels of community involvement are much more likely to rent land. This indicates that land rental is embedded within a set of community-wide social institutions and thus the social capital of community involvement is an asset to a household in this area.

The relationships between household labor and nonredundant ties, on the one hand, and rental of land, on the other, however, are not linear. Households with middle-level household labor and middle-level helping networks, operationalized in terms of numbers of nonredundant ties, were the most likely to rent land. This shows that merely accumulating adults in a household or merely adding persons to one's social exchange helping networks will not necessarily improve the economic situation of a peasant household. Rather, households with stable social structures, especially inter-generational households that consist of a married couple, their children, and other adult members seem to be especially advantaged.

The data also show that there has been some significant movement of households in these villages to purchase or trade for various types of lower cost agricultural equipment, other than tractors. In addition, the panel surveys show that a substantial number of households purchased automobiles that are used to transport, grain, livestock and inputs to and from fields and markets.

Peasant households in Vengerovka own fewer mechanical tools than peasant households in Latonovo or Bolshoe Sviattsovo. Therefore, we might expect that the larger amount of rented land in Vengerovka and the higher levels of household production in that village, which will be discussed in the next chapter, will produce higher levels of stress for households in Vengerovka, compared to households in Latonovo or Bolshoe Sviattsovo.

Household labor and community involvement is positively associated with a household's acquisition of mechanical agricultural tools.

The ability of a household to care for livestock is one of the most important factors in determining the economic well being of its members. During the three years of the panel study there was a significant overall increase in the number of hogs and poultry which households maintained. There were, however, substantial differences between villages, with households in Vengerovka having considerably more livestock than households in the other two villages.

The clearest indications of the importance of human and social capital in giving relative advantages to some households over others is seen in the relationship between household labor and community involvement, on the one hand, and the amount of livestock cared for by the household on the other. Households with more household labor are able to devote more effort to the labor-intensive activities associated with caring for pigs and poultry.

The strong positive relationship between community involvement and how much livestock a household keeps illustrates again how economic success in the Russian village is dependent to some extent on the social capital of the household. Since some livestock is kept on common land, and neighbors need to watch over animals when household members are not in the village, those households that have maintained their social connections with their neighbors have a clear advantage over other households that have been less involved in the village.

At this juncture, we will turn to the actual production and sale of agricultural products by the households in the panel study. Here we will see how initial household advantages in human and social capital, reinforced by the relationship between those advantages and access to physical capital, serve to differentiate the ability of households to compete in the emerging market economy.

8 Household Agricultural Production and Sales

David J. O'Brien, Valeri V. Patsiorkovski and Larry D. Dershem

Introduction

In this chapter, we will examine the production and sales of peasant households in the three villages during the three years of the panel study. Our primary focus will be to identify the relationships between the types of household capital (human, social, and physical) described in earlier chapters and the ability of households to produce and sell various types of commodities.

Although peasant households were allowed to sell a portion of what they produced on their private plots during the Soviet period, the opportunities for increasing production and sales increased significantly during the early 1990s. As shown in Figures 1.1 and 1.2, peasant household production is a significant contributor to overall production in Russian agriculture, and, in certain commodity areas, especially meat and dairy, has increased its share of production vis-à-vis the large agricultural enterprises. Our central thesis is that the ability of a household to take advantage of these opportunities for increased production and sales will be associated with the level of its human (household labor), social (exchange helping networks and community attachment) and physical (access to tools, rental of land, and amount of livestock) capital.

Peasant households produce a variety of commodities for consumption and for sale. The former, which is a type of nonmonetized income, will be examined in relation to monetized income in the next chapter. Household sales of commodities produced on peasant plots and rented land are a critical source of monetized income for peasant households. This income may be used to buy finished goods, such as clothing or housewares or contribute to the purchase of more expensive durable goods, such as a VCR, refrigerator, or even an automobile.

Because household production and sales have such a critical bearing on both monetized and nonmonetized income, they also have a critical bearing on the quality of life of the Russian peasant household. As far as nonmonetized income is concerned, the ability of a household to produce meat, milk, potatoes, vegetables, and fruit will directly affect the quality of the food that members of that household eat. Monetized income, resulting from sales of a household's production will directly affect the ability of the household to purchase the more expensive, often imported, goods and services that have begun to appear in shops in regional centers and large cities. The ability of a household to pay for these goods with money gained from sales of pork or poultry or sour cream will determine, to a significant degree, whether the household is able to experience any of the material benefits of a new market economy. In short, the amount of agricultural products, whether they are consumed by the household, or sold to purchase finished goods or services, are a major source of differentiation between households in the post-Soviet Russian villages. Thus, understanding the factors that produce differences in the ability of households to produce and sell agricultural commodities is critical in understanding the emerging system of stratification in Russian villages.

Peasant household production is extremely labor-intensive. Although increasingly peasant households have obtained some types of mechanical equipment, their ability to maintain plots and care for animals depends on their ability to obtain household labor and to build personal helping networks. At the same time, we expect that a critical factor in differentiating between households will be the extent to which they are involved in the life of their village communities. The expectation here is that even though overall community involvement has declined in Russian villages in recent years, those households that maintain high levels of involvement will have a significant social capital advantage that will translate into lower transaction costs related to marketing and thus result in increased sales of their products. Finally, our expectation is that the various types of human and social capital also will operate indirectly on agricultural production and sales through their association with the ability of the household to access physical capital. It is expected that human and social capital advantages will multiply as they lead to increased opportunities for households to gain access to equipment, land and animals, thereby further differentiating between households in the villages.

Household Production by Village from 1995 to 1997

The mean level of production, reported in kilograms, of various commodities and the total weighted average of all commodities, by peasant households in the three villages during the three years of the panel study is shown in Table 8.1. The weighting procedure, used to measure total household production, is based on summaries of the 1995 market value of each commodity (see Chapter 3).

The production of all of the commodities listed, with the exception of milk, increased significantly from 1995 to 1997. The most important increases were in the production of meat and potatoes. The increase in production of potatoes is associated primarily with their use as feed for livestock, especially in winter. Potatoes are easy to store and peasant households cook them in the winter and feed them to pigs and poultry. This is a very arduous process that requires substantial hand labor.

There are significant differences in production of all commodities between villages during each year of the study. The most important differences are in production of meat and potatoes. Households in Vengerovka have higher levels of meat production than do households in Bolshoe Sviattsovo for all three years and higher levels of meat production than do households in Latonovo in 1995 and 1996.

The most important figures are those pertaining to total weighted household agricultural production (Table 8.2). These figures are adjusted to account for the values of different commodities - e.g., a kilogram of meat has a value six times greater than a kilogram of potatoes - whether they are consumed by the household, sold for cash, or traded for goods or services. Thus, this weighted figure is a good indicator of a household's ability to obtain both nonmonetary (i.e., consumption) and monetary (sales) advantages from production. From 1995 to 1997, households in the sample increased their production by 23.3 percent, which is a substantial increase in a three year period.

These figures show, however, substantial differences between villages in agricultural production. Households in Latonovo, on average, have much lower rates of production than do households in either Vengerovka or Bolshoe Sviattsovo. The conditions for growing grain and potatoes, which are used to feed animals, are similar in Latonovo and Vengerovka. Both villages are located in the *chernozem* zone. It is especially striking that household production in Latonovo is much less than in Bolshoe Sviattsovo, which is located outside of the black earth zone and thus has a substantial disadvantage in growing conditions vis-à-vis the other two areas.

Household Agricultural Production and Sales 135

Table 8.1: Mean Household Production of Meat and Milk (kilograms) in Three Russian Villages from 1995 to 1997

Village	Meat 95	Meat 96	Meat 97	Milk 95	Milk 96	Milk 97	Vegetables 95	Vegetables 96	Vegetables 97	Potatoes 95	Potatoes 96	Potatoes 97	Fruit 95	Fruit 96	Fruit 97
Latonovo (n=157)	185	221	325	1860	1917	1857	123	124	129	807	710	535	20	27	36
Vengerovka (n=156)	322	312	393	2304	2780	2897	230	379	352	2360	2605	3233	45	41	55
Sviattsovo (n=150)	168	164	249	2780	2666	3078	266	300	435	1694	1806	2019	61	40	91
Total (N=463)	226	233	323	2308	2308	2603	206	267	303	1618	1704	1749	206	267	303

Meat production by year F(2)=18.56, p<.001. Meat production by village 1995: F(2)=20.23, p<.001; Scheffe, Vengerovka x Latonovo & Sviattsovo, p<.001. Meat production by village 1996: F(2)=12.83, p<.001; Scheffe, Vengerovka x Latonovo, p<.01, Vengerovka x Sviattsovo, p<.001. Meat production by village 1997: F(2)=8.92, p<.001; Scheffe, Vengerovka x Sviattsovo, p<.001.
Milk production by year F(2)=1.570, p=n.s. Milk production by village 1995: F(2)=5.895, p<.001; Scheffe, Latonovo x Sviattsovo, p<.01. Milk production by Village 1996: F(2)=5.44, p<.01; Scheffe, Latonovo x Vengerovka, p<.01, Latonovo x Sviattsovo, p<.05. Milk production by Village 1997: F(2)=9.68, p<.001; Scheffe, Latonovo x Vengerovka, p<.01, Latonovo x Sviattsovo, p<.001.
Vegetable production by year F(2)=4.03, p<.05. Vegetable production by village 1995: F(2)=5.97, p<.01; Scheffe, Vengerovka, p<.05, Latonovo x Sviattsovo, p<.01. Vegetable production by village 1996: F(2)=6.91, p<.001; Scheffe, Latonovo x Vengerovka & Sviattsovo, p<.001, Latonovo x Sviattsovo, p<.05. Vegetable production by village 1997: F(2)=13.61, p<.001; Scheffe, Latonovo x Vengerovka & Sviattsovo, p<.001.
Potato production by year F(2)=2.13, p=n.s. Potato production by village 1995: F(2)=28.08, p<.001; Scheffe, Latonovo x Vengerovka & Sviattsovo, p<.001; Vengerovka x Sviattsovo, p<.01. Potato production by village 1996: F(2)=21.24, p<.001; Scheffe, Latonovo x Vengerovka & Sviattsovo, p<.001, Vengerovka x Sviattsovo, p<.05. Potato production by village 1997: F(2)=68.49, p<.001; Latonovo x Vengerovka & Sviattsovo, p<.001, Vengerovka x Sviattsovo, p<.001.
Fruit production by village: By Year F(2)=12.52, p<.001. By Village 1995: F(2)=6.31, p<.01. By Village 1996: F(2)=4.63, p<.01. By Village 1997: F(2)=28.07, p<.001

In short, the lower level of household production in Latonovo cannot be attributed to disadvantages in natural conditions, like soil or climate. Moreover, although it might be argued that Bolshoe Sviattsovo has an advantage over Latonovo with respect to proximity to large urban markets, a similar advantage cannot be found in comparing Vengerovka to Latonovo. Both of the latter villages are located in equal proximity to urban markets. The explanation for the relative disadvantage of Latonovo, which will be shown later, is that this village does not possess a mechanism with which households can rent land to increase their productive capacity. This provides some support for hypotheses 12a that villages which provide structural support for the development of human and social capital will create conditions that result in higher levels of production in peasant households (*Agronpromyshlennyi kompleks Belgorodskoi oblasti sostoianie i perspektivy* 1998). Further insight into these relationships will be found in our discussion of the impact of rental of land on household production that is described later in this chapter.

Table 8.2: Mean Total Weighted Household Agricultural Production by Village

Village	1995	1996	1997
Latonovo (n=157)	4,849	5,060	5,747
Vengerovka (n=156)	8,020	9,065	10,340
Sviattsovo (n=150)	7,201	7,131	8,653
Total (N=463)	6,679	7080	8,236

By Year (F2)=7.46, p<.001. By Village 1995: F(2)=14.95, p<.001; Scheffe, Latonovo x Vengerovka & Sviattsovo, p<.001. By Village 1996: F(2)=15.66, p<.001; Scheffe, Latonovo x Vengerovka, p<.001, Latonovo x Sviattsovo, p<.01. By Village 1997: F(2)=18.60, p<.001; Scheffe, Latonovo x Vengerovka & Sviattsovo, p<.001.

Effect of Human and Social Capital on Household Production

The mean levels of household production within different levels of household labor are shown in Table 8.3. The strong positive relationship between level of household labor and level of household production supports hypothesis 1a. In labor-intensive agriculture, households with

Household Agricultural Production and Sales 137

more available labor have a clear advantage over households with less labor.

It is important to remember that this household labor is *embedded* within household social capital. As shown earlier in Chapter 4, the value of the variable "weighted number of adults" is strongly associated with the structure of the household. Production in this type of household enterprise requires cooperation between household members and thus individual members are contributing to a "public good" (Olson 1971). It is the *trust* and cooperation produced by long-term social relationships between household members that forms the social capital within which household labor produces agricultural commodities. The highest level of labor is found in households comprised of husbands and wives, children and other adults. The lowest level of labor is found in households consisting of single-adults, usually elderly persons, retired couples or single parents (see Chapter 5).

Table 8.3: Mean Household Agricultural Production (kilograms) by Household Labor from 1995 to 1997

Weighted Number of Adults	1995	1996	1997
0-1.74	2,706 (n=143)	2,895 (n=146)	3,338 (n=147)
1.75-2.74	7,803 (n=164)	7,859 (n=177)	9,442 (n=173)
2.75-3.74	8,767 (n=108)	8,920 (n=95)	10,990 (n=93)
3.75-4.74	9,710 (n=34)	12,267 (n=35)	12,578 (n=44)
4.75 +	10,647 (n=14)	23,050 (n=10)	18,940 (n=6)
Total (N=463)	6,679	6,530	8,236

1995: $F(4)=38.12$; Scheffe, 1 x 2, 3 & 4, $p<.001$; 1996: $F(4)=7.81$; Scheffe, 1 x 2 & 3, $p<.05$, 1 x 4, $p<.001$; 1997: $F(4)=41.43$, $p<.001$; Scheffe 1 x 2, 3, 4 & 5, $p<.001$, 2 x 4, $p<.05$, 2 x 5, $p<.01$.

Graph 8.1 shows the predicted value of each additional unit of household labor for total household production. Overall, the addition of a working age adult, age 18-64, adds 2,792 kilos to the predicted amount of production by a household. The addition of a child 12 to 14 years of age would result in an expected gain of approximately 1,500 kilograms to a household's production in a single year. The advantage of additional household members begins to decline after adding the equivalent of a third adult to the household.

Graph 8.1: Predicted Amount of Agricultural Production (kilograms) by Weighted Number of Adults

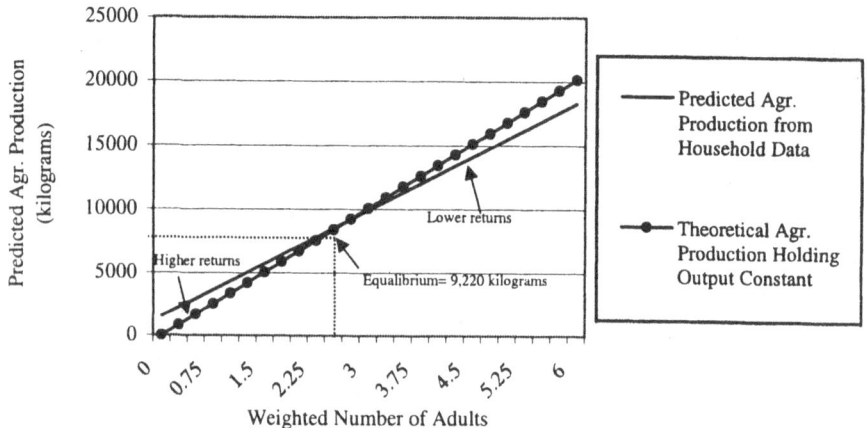

One interpretation of the diminishing returns of additional household labor is the "free-rider" effect. In our view, a more plausible interpretation is that external and internal conditions limit the amount of time that additional household members can devote to agricultural production. Externally, households have limitations related to land, technology and marketing. Internally, there is a limit to the amount of food that households can consume or preserve. Moreover, households are not purely production units; household members must devote time to other necessities such as employment and household tasks such as care of children.

Household Agricultural Production and Sales 139

Nevertheless, the diminishing returns of additional adults are quite modest, showing that the overall relationship between household labor and household production is linear and quite strong.

Table 8.4: Mean Weighted Production (kilograms) by Number of Non-Redundant Social Network Ties

Number of Ties	1995	1996	1997
0-3	4,735 (n=102)	4,538 (n=91)	5,688 (n=89)
4-5	6,007 (n=155)	6,909 (n=145)	7,376 (n=133)
6	8,026 (n=66)	7,386 (n=74)	9,489 (n=79)
7-10	8,184 (n=122)	8,934 (n=139)	9,855 (n=144)
11-17	3,439 (n=18)	5,364 (n=14)	8,735 (n=18)
Total (n=463)	6,679	7,080	8,236

1995: $F(3)=7.95$, $p<.001$; Scheffe, 1 x 3, $p<.01$, 1 x 4, $p<.001$, 2 x 4, $p<.05$. 1996: $F(3)=6.89$, $p<.001$; Scheffe, 1 x 4, $p<.001$. 1997: $F(3)=8.30$, $p<.001$; Scheffe, 1 x 3, $p<.01$, 3 x 4, $p<.001$, 2 x 4, $p<.05$.

Table 8.4 shows the relationship between different sizes of social networks on a household's production. These findings provide mixed support for hypothesis 2a, showing that the relationship between social networks and production is complex.

The measure of social network size in this instance is the number of non-redundant ties in the total number of specific use networks (six networks in all) possessed by each household. This is an indicator of the total number of persons who help a household with all of the tasks it needs to accomplish. As shown in Chapter 5, there has been a substantial increase in the number of different tasks to which each person in the overall helping networks of a household contributes. There has not been, however, any increase in the total number of persons who are helping a household. In short, the same number of persons is doing more things to help a household today than they were doing in earlier times. This means that there has been an increase in redundant, but not in non-redundant ties.

This specific adjustment to the increased demands of the market economy appears to be very efficient. Graph 8.2 shows that there is a curvilinear relationship between the number of persons in a household's helping network and its ability to produce agricultural commodities. From 0

140 *Household Capital and the Agrarian Problem in Russia*

to 10 persons, each person added to a household's networks increases production by an average of 1,700 kilograms. After 10 persons, however, additional persons in the network are associated with a decline in production, which results, on average, in a loss of 87 kilograms for each person added.

Graph 8.2: Number of Nonredundant Ties and Predicted Household Production (kilograms)

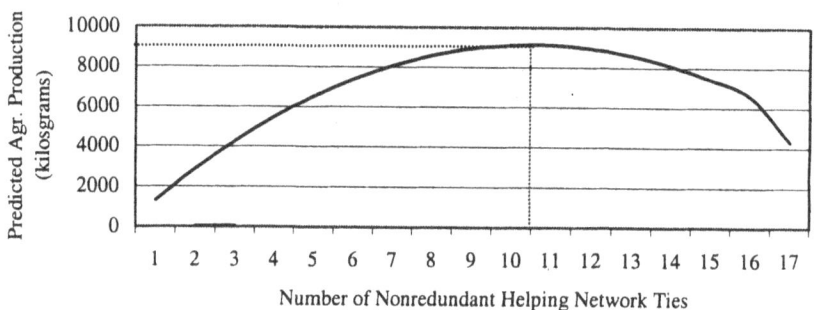

Number of Nonredundant Helping Network Ties

This curvilinear relationship is caused by households with moderately high levels of nonredundant tie networks being quite different than households with extremely large numbers of persons in their networks (as shown in Chapter 5). Moderate size helping networks, from seven to ten persons, are most likely to be from demographic types four and five (husband and wife with children and husband and wife with children and other adults). Households with extremely large size of helping networks, eleven or more, consist of single persons (demographic type 1) or older retired couples (demographic type 2). The networks of the former are more likely to be part of an ongoing household agricultural production and sales system, whereas the latter networks are more likely to consist of family members and neighbors who assist vulnerable persons in times of need. Thus, it is not surprising that households with moderate size of social networks (7 to 10 persons) have higher levels of production than do households with more than 10 nonredundant ties.

The relationship between community involvement and household production, as suggested by hypothesis 3a will be discussed later.

Effect of Physical Capital on Household Production

It will be recalled from Chapter 1 that a critical intervening set of variables in the process of household production and sales will be those pertaining to physical capital. Hypothesis 6a suggests that household human and social capital will be associated with the ability of households to acquire physical capital. The physical capital variables *multiply* advantages or disadvantages for households with various levels of human and social capital. The data from the panel survey provide support for hypothesis 6a.

One type of physical capital that has a substantial effect on the ability of a household to produce agricultural commodities is the rental of land. The impasse over land reform at the federal level has severely limited the transfer of land and the creation of a true land market. Nonetheless, rental arrangements have been taking place in rural areas of Russia. Typically, the amount of land which is rented is not very large; usually less than a hectare is involved in any single transaction (See Chapter 6; see also, Wegren 1998:18-19). Nonetheless, the addition of even a small plot of land can make a significant difference in the output of labor intensive agriculture, especially when it involves the production of a value-added product, such as hams, sausages or sour cream. Households with more household labor, social networks and community attachments will have an advantage in this regard. They will have additional hands that can work rented land and social networks and community attachments, which can lead to the development of social relationships upon which such rental arrangements are based. The villages of Vengerovka and Bolshoe Sviattsovo have mechanisms for households to rent land while Latonovo does not.

Rental of land is strongly associated with a household's capacity to produce agricultural commodities (see Table 8.5). Households with rented land produced, on average, 80 percent more than other households did in 1995 and that advantage increased to 84 percent in 1996 and 93 percent in 1997. Moreover, the opportunity to rent land for household use through formal agreements that are present in Vengerovka and Bolshoe Sviattsovo accounts for the higher average levels of production in these two villages than in Latonovo. These findings provide support for hypothesis 12a.

142 *Household Capital and the Agrarian Problem in Russia*

Table 8.5: Mean Total Household Agricultural Production (kilograms) by Rental of Land

Rented land	1995	1996	1997
No	5,583 (n=349)	5,947 (n=358)	6,565 (n=336)
Yes	10,037 (n=114)	10,946 (n=105)	12,657 (n=127)
Total (N=463)	6,679	7,080	8,236

1995:F(1)=64.54, p<.001; 1996:F(1)=52.95, p<.001.; 1997:F(1)=82.27, p<.001.

The most labor-intensive aspect of peasant household production in Russian villages is the care of animals. This is also the type of physical capital that will have the greatest effect on enhancing advantages that a household possesses because of its human and social capital. The number of animals a household can keep is directly proportional to the labor it can devote to their maintenance. Cattle, hogs, sheep and fowl have to be fed and sheltered on a daily basis. This involves moving them from household sheds and barns to fields and ponds each day, returning them to the household at night. In the case of cows, this involves milking twice a day. All of these activities are especially difficult during the Russian winter, in which feed often has to be heated. Thus, households with more working-age adults can be expected to have an easier time handling a larger number of animals. In addition, having extensive social networks and community attachments will increase the number of persons outside of the household itself that a household can depend on to help care for their animals.

Table 8.6: Mean Total Household Agricultural Production (kilograms) by Weighted Number of Animals

Weighted Number of Animals	1995	1996	1997
0-50	1,456 (n=114)	1,293 (n=114)	1,702 (n=120)
51-325	4,765 (n=113)	5,124 (n=122)	5,899 (n=114)
326-611	8,363 (n=131)	8,911 (n=113)	10,131 (n=102)
612 +	12,309 (n=105)	13,148 (n=114)	14,986 (n=127)
Total (N=463)	6,679	7,080	8,236

1995:$F(3)=165.93$, $p<.001$; Scheffe, 1 x 2, 3 & 4, $p<.001$, 2 x 3 & 4, $p<.001$, 3 x 4, $p<.001$;
1996:$F(3)=125.06$, $p<.001$; Scheffe 1 x 2, 3 & 4, $p<.001$, 2 x 3 & 4, $p<.001$, 3 x 4, $p<.001$;
1997:$F(3)=170.39$, $p<.001$; Scheffe 1 x 2, 3 & 4, $p<.001$, 2 x 3 & 4, $p<.001$.

Table 8.6 shows a strong relationship between the number of animals possessed by a household and its overall level of agricultural production. The weighted figures for number of animals were generated by the procedures described in Chapter 3 that give different weights for different types of animals (e. g., a chicken is "1", a pig is "50" and a cow is "250"). These figures also are consistent with the niche in Russian agriculture that is occupied by peasant households. As will be recalled from Table 2.1 in Chapter 2, peasant households account for more than half of all meat produced in Russia in 1996. Moreover, these households also account for almost half (45.4 percent) of all dairy products, which also are dependent on animals, produced in Russia.

Modeling the Effect of Household Capital on Production

The results of the structural equation modeling of the pooled panel data, using the AMOS program (Arbuckle 1997) is presented in Figure 8.1. The model shows the effects of different types of household capital on peasant household production over time. Since the data from all three years of the panel survey are pooled, the findings show which variables have the most stable and long-term effects on household production. The Chi-Square value showing a nonsignificant relationship means that the model has a high degree of fit between what is observed empirically and our theoretical model. Because of a small number of extremely high values a logarithmic transformation was used for the variables weighted number of animals and total weighted production.

There is strong support for hypothesis 1a that households with more labor are at a considerable advantage vis-à-vis other households with respect to production capacity. Earlier we saw that the association between household labor and household production was very strong (see Table 3). The direct association between household labor and household production vanishes, however, when all of the other social and physical capital variables are included in the structural equation models. Thus there is no support for hypothesis 4a that there would be a direct effect of household labor on household production. *All of the effects of household labor are indirect, operating through other types of household capital.* There is, however, strong support for hypotheses 5a, 6a and 7a that household labor has an indirect effect on household production through its association with access to mechanical equipment, rental of land and the number of animals possessed by the household. The total standardized effect of household labor (0.18) that operates through these three types of physical capital (i.e., multiplying the indirect paths for each type of physical capital and then summing them) is three times greater than the total effect of land (0.06) on household production.

As in the case of peasant smallholder farming (Netting 1993) in third world countries, Russian peasant households with more hand-labor have a clear advantage over their neighbors with less labor. This additional labor, however, cannot merely be calculated in the simple additive sense described by Chaianov (1966), and others (Deere and de Janvry 1981). In the Russian peasant household economy there are elements of the traditional moral economy of the peasant household, but also there are some types of physical capital that are part of a more technologically advanced market-driven agricultural economy. Thus, households with more working-hands are able to purchase mechanical equipment and rent land.

Figure 8.1: Structural Equation Model of the Combined Effects of Different Types of Household Capital on Peasant Household Agricultural Production ($X^2=11.33$, $df=10$, $p=.33$, $GFI=.998$, $AGFI=.992$)

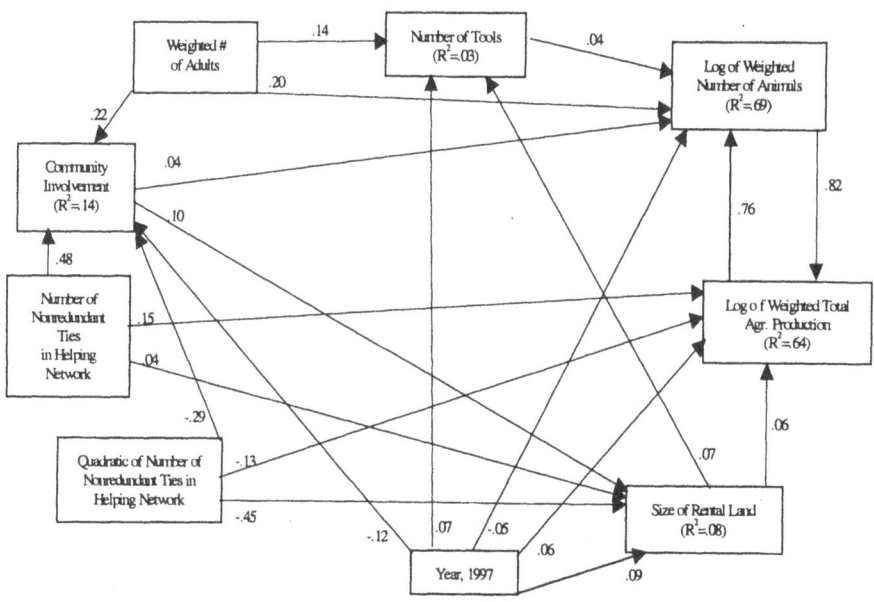

146 Household Capital and the Agrarian Problem in Russia

Table 8.7: Standardized Regression Coefficients for Observed Exogenous Variables and Explained Variance (R^2) for Structural Model of Peasant Household Agricultural Production

Observed endogenous	R^2	Standardized Regression weights	Observed exogenous
Log of Weighted Agricultural Production	0.64		
		0.82	Log of Weighted Number of Animals
		0.15	Number of Nonredundant Ties in Helping Network
		-0.13	Quadratic of Number of Nonredundant Ties in Helping Network
		0.06	Size of Rental Land
		0.06	Year 1997
Log of Weighted Number of Animals	0.69		
		0.76	Log of Weighted Total Agricultural Production
		0.20	Weighted Number of Adults
		0.06	Year 1997
		0.04	Community Involvement
		0.04	Number of Tools
Community Involvement	0.14		
		0.48	Number of Nonredundant Ties in Helping Network
		-0.29	Quadratic of Number of Nonredundant Ties in Helping Network
		0.22	Weighted Number of Adults
		-0.12	Year 1997
Size of Rental Land	0.08		
		0.40	Number of Nonredundant Ties in Helping Network
		-0.45	Quadratic of Number of Nonredundant Ties in Helping Network
		0.10	Community Involvement
		0.10	Year 1997
Number of Agricultural Tools	0.03		
		0.14	Weighted Number of Adults
		0.07	Size of Rental Land
		0.07	Year 1997

$X^2=11.33$, df=10, p=.33, GFI=.998, AGFI=.992

Both of these types of physical capital are not as readily available in a third world peasant economy.

Household Agricultural Production and Sales 147

Again, it is important to recall the point made in Chapter 6 that the opportunity for peasant households to rent land depends on support from the local village administration. In Vengerovka and Bolshoe Sviattsovo the local village administration has supported the development of a procedure by which households can rent land and thereby increase the amount of land they cultivate. Alternatively, in Latonovo, with its more conservative administration, this opportunity was never developed and, as a result, households have not been able to cultivate as much land and, in turn, mean levels of production are much lower in this village than in the other two.

The model shows that the strongest path through which household labor affects production is very typical of traditional household production; namely, the ability of households with more hand labor to feed and otherwise care for larger numbers of animals.

There is a moderate zero-order correlation between household labor and the number of persons in a household's personal helping networks (Pearson r = 0.30). When the non-linear quadratic term for nonredundant personal helping network is added to the structural equation model, the causal path between household labor and number of nonredundant ties drops out and is replaced by a covariance between the two variables. Substantively, this means that although there is a relationship between household labor and personal network ties, it is not exceptionally strong. In turn, this means that personal helping network ties, which is a purer form of social capital, does stand on its own, as a source of higher levels of household production.

The structural equation model, as well as Table 8.7, indicates that the relationship between the size of a household's helping network and its level of production is more complex than had been expected. Hypothesis 2a had proposed that households with larger social networks would have an advantage over other households in agricultural production. The structural equation model shows, however, that moderately large networks, in the range of 7-10 persons, create the greatest advantages for peasant households. The non-linear quadratic term measures the declining production of those households that have exceptionally high numbers (11 to 17) of persons in their helping networks. We suggested earlier that typically these households consist only of elderly persons.

The number of persons in the household's helping network affects production in two ways. First, there are two direct paths from the number of nonredundant ties to production. The additive term, reflecting the effect of the number of ties from 0 to 10 is positive and strong, while the quadratic term, reflecting the effect of networks sized 11 to 17 is negative and also strong. Second, the number of nonredundant ties has a positive relationship with renting land and thus through this path has an indirect positive effect on household production.

There is mixed support for hypothesis 3a that had stated that households that were more integrated into village life would have higher levels of productivity. The measure of integration into village life, the community involvement index, does not have a direct effect on agricultural production but it does have an indirect effect through its association with rental of land and number of animals owned by the household.

The negative path from year 1997 to community involvement reflects the overall decline in community involvement from 1996 to 1997 that was shown in 3. Finally, the model shows that among the three years, when controlling for all variables, important changes occurred in 1997. In that year there was a significant increase in agricultural production and a corresponding decline in the number of animals in the household. This indicates that households were slaughtering more livestock during that year.

Effect of Human and Social Capital on Household Sales

The sales of various commodities by households in the three villages from 1995 to 1997 are shown in Table 8.8 The proportion of sales to total household production varies considerably from one commodity to another. The proportion of meat sold to meat produced ranges from 47.3 percent in 1995 to 58.2 percent in 1997. The proportion of sales to production of dairy products is even higher, ranging from 73.9 percent in 1995 to 90.8 percent in 1997. A much smaller proportion of vegetables and fruit produced by peasant households are sold, ranging from 12.1 percent to 22.8 percent for vegetables in the total sample and, in the case of fruit, in some villages no fruit was sold. These figures illustrate at the micro-level what was shown earlier in Table 2.1 that peasant households are increasingly specializing in the sale of meat and dairy products.

Table 8.8: Mean Total Weighted Household Sales of Different Commodities (kilograms) by Three Russian Villages from 1995 to 1997

Village	Meat			Dairy (milk)		
	1995	1996	1997	1995	1996	1997
Latonovo	81 (n=123)	109 (n=137)	207 (n=144)	1858 (n=85)	1405 (n=87)	1325 (n=89)
Vengerovka	164 (n=140)	165 (n=146)	212 (n=155)	2122 (n=96)	2337 (n=95)	3022 (n=94)
Sviattsovo	65 (n=110)	85 (n=107)	136 (n=128)	1130 (n=92)	998 (n=90)	2710 (n=88)
Total	107 (N=373)	123 (N=390)	188 (N=427)	1705 (N=273)	1596 (N=272)	2363 (N=271)

Meat Sales by Year $F(2)=15.75$, $p<.001$. Meat Sales by Village 1995: $F(2)=10.98$, $p<.001$; Scheffe, Latonovo x Vengerovka & Sviattsovo, $p<.001$. Meat Sales by Village 1996: $F(2)=22.79$, $p<.001$; Scheffe, Vengerovka x Latonovo & Sviattsovo, $p<.001$. Meat Sales by Village 1997:
Milk Sales by Year $F(2)=15.28$, $p<.001$. Milk Sales by Village 1995: $F(2)=11.90$, $p<.001$. Scheffe, Latonovo compared to Sviattsovo, $p<.001$. Milk Sales by Village 1996: $F(2)=17.25$, $p<.001$. Scheffe, Latonovo compared to Vengerovka, $p<.001$, Vengerovka compared to Sviattsovo, $p<.001$. Milk Sales by Village 1997: $F(2)=22.43$, $p<.001$. Scheffe, Latonovo compared to Vengerovka and Sviattsovo, $p<.001$.

Table 8.8 continued

Village	Vegetables			Potatoes			Fruit		
	1995	1996	1997	1995	1996	1997	1995	1996	1997
Latonovo	0	0	0	248	195	178	0	0	0
	(n=107)	(n=140)	(n=143)	(n=157)	(n=157)	(n=157)	(n=61)	(n=69)	(n=90)
Vengerovka	20	102	51	469	554	1278	0	0	0
	(n=133)	(n=154)	(n=153)	(n=156)	(n=156)	(n=156)	(n=68)	(n=97)	(n=114)
Sviattsovo	48	75	83	468	196	285	4	1	5
	(n=144)	(n=145)	(n=148)	(n=150)	(n=150)	(n=150)	(n=90)	(n=99)	(n=118)
Total Sales	25	61	45	393	316	583	25	61	45
	(N=384)	(N=439)	(N=444)	(N=463)	(N=463)	(N=463)	(N=384)	(N=439)	(N=444)

Vegetable Sales by Year F(2)=.83, p=n.s. Vegetable Sales by Village 1995: F(2)=1.01, p=n.s. Vegetable Sales by Village 1996: F(2)=1.87, p=n.s. Vegetable Sales by Village 1997: F(2)=1.48, p=n.s.
Potato Sales by Year F(2)=4.17, p<.05. Potato Sales by Village 1995: F(2)=.85, p=n.s. Potato Sales by Village 1996: F(2)=5.45, p<.01. Scheffe, Latonovo compared to Vengerovka, p<.05, Vengerovka compared to Sviattsovo, p<.05. Potato Sales by Village 1997: F(2)=31.73, p<.001. Latonovo compared to Vengerovka, p<.001, Vengerovka compared to Sviattsovo, p<.001.
Fruit Sales by Year F(2)=.94, p=n.s. Fruit Sales by Village 1995: F(2)=1.25, p=n.s. Sales by Village 1996: F(2)=.837, p=n.s. Sales by Village 1997: F(2)=2.636, p=n. s.

Household Agricultural Production and Sales 151

Meat and milk sales increased significantly over the three years of the panel study. Moreover, village differences, which were shown earlier with respect to production are even more striking with respect to sales. Both Vengerovka and Bolshoe Sviattsovo have higher average levels of sales than does Latonovo. Sales, are higher in Vengerovka than in Bolshoe Sviattsovo (see Table 8.9). The higher level of sales in Vengerovka than in the other two villages reflects two conditions. First, as noted earlier, peasant households in Vengerovka have local government support for rental of land to increase the size of their plots. Second, a credit scheme, to assist peasant households to obtain money to improve their facilities for production, storage, and processing of agricultural products has been in place in Belgorod *Oblast* since 1994. No comparable plan exists in the other two villages.

The "fund for the support of individual buildings in rural areas" (*fond podderzhki individualnogo zhilishchnogo stroitelstva na sele*) encourages households to borrow money for capital improvements but allows them to repay these loans with the meat and dairy products they produce. This program appears to have created very powerful incentives that encourage households in Vengerovka to produce more commodities for sale (O'Brien et al. 1998a:44-45). These findings provide strong support for hypotheses 12b.

Table 8.9: Total Household Agricultural Sales (kilograms) by Village and Year

Village	1995	1996	1997
Latonovo (n=157)	1,968	1,761	2,283
Vengerovka (n=156)	3,269	3,714	5,270
Sviattsovo (n=150)	1,721	1,470	3,335
Total (N=463)	2,326	2,325	3,630

Total Weighted Sales by Year: $F(2)=20.50$, $p<.001$. Total Weighted Sales by Village 1995:10.98, $p<.001$. Scheffe, Vengerovka compared to Latonovo and Sviattsovo, $p<.001$. Total Weighted Sales by Village 1996: $F(2)=22.79$, $p<.001$. Scheffe, Vengerovka compared to Latonovo and Sviattsovo, $p<.001$. Total Weighted Sales by Village 1997: $F(2)=22.77$, $p<.001$. Scheffe, Vengerovka compared to Latonovo and Sviattsovo, $p<.001$.

152 Household Capital and the Agrarian Problem in Russia

Table 8.10 shows the levels of household sales by different levels of household labor. These findings provide strong support for hypothesis 1b. The greatest gains in sales are made when we move from the first to the second levels of household labor, where there is an increase of almost 4 times in the amount of weighted sales. The amount of sales in the third level of household labor is slightly less than a 7 percent increase over the second level, while the increase in the fourth level is 15 percent. Households with the highest level of labor have almost a third more sales than the next highest level and almost 7 times more sales than do households with the lowest level of household labor.

Table 8.10: Mean Total Weighted Sales (kilograms) by Household Labor from 1995 to 1997

Weighted Number of Adults	1995	1996	1997
0-1.74	590 (n=143)	774 (n=146)	1,011 (n=147)
1.75-2.74	2,887 (n=164)	2,542 (n=177)	4,410 (n=173)
2.75-3.74	3,084 (n=108)	2,836 (n=95)	5,172 (n=93)
3.75-4.74	3,546 (n=34)	5,409 (n=35)	5,418 (n=44)
4.75 +	4,686 (n=14)	5,472 (n=10)	8,337 (n=6)
Total (N=463)	2,326	2,325	3,630

1995: $F(4)=19.20$, $p<.001$. Scheffe, 1 x 2 through 5, $p<.001$; 1996: $F(4)=21.76$, $p<.001$. Scheffe, 1 x 2 through 5, $p<.001$, 2 x 4, $p<.001$, 3 x to 4, $p<.001$; 1997: $F(4)=28.98$, $p<.001$. Scheffe, 1 x 2 through 5, $p<.001$, 2 x 4 & 5, $p<.05$.

Table 8.11: Mean Total Weighted Sales (kilograms) by Number of Non-Redundant Ties

Nonredundant Ties	1995	1996	1997
0-3	1,323 (n=102)	1,189 (n=91)	2,232 (n=89)
4-5	1,828 (n=155)	2,178 (n=145)	3,157 (n=133)
6	2,994 (n=66)	2,799 (n=74)	4,681 (n=79)
7-10	3,294 (n=122)	3,082 (n=139)	4,436 (n=144)
11-17	3,439 (n=18)	1,191 (n=14)	2,981 (n=18)
Total (N=463)	2,326	2,325	3,630

1995:$F(3)=7.88$, $p<.001$. Scheffe, 1 x 3, $p<.05$, 1 x 4, $p<.001$, 2 x 4, $p<.01$; 1996:$F(3)=5.52$, $p<.001$. Scheffe, 1 x 3, $p<.05$, 1 x 4, $p<.001$; 1997:$F(3)=5.94$, $p<.001$. Scheffe, 1 x 3, $p<.01$, 1 x 4, $p<.01$.

Again, it is important to remember that this household labor is *embedded* within household social capital (see Chapter 5). These findings, as well as the findings shown earlier with respect to household production, show clearly that the amount of social capital within the household is a very powerful source of differentiation between households in the Russian countryside.

The average amount of sales by households with different size helping networks, the latter being measured by the number of non-redundant ties in a household's networks, is shown in Table 8.11. As in the case of household production (see Table 8.3) there is mixed support for hypothesis 2b. There is a curvilinear relationship between the number of ties in a household's helping network and its level of sales. There is an increasing level of sales up to the moderately high level of 7 to 10 people, but sales begin to decline when households reach a total number of nonredundant ties of 11 or more. The moderately high level of personal networks are associated with younger families with children and younger families with children and other adults, whereas the extremely large networks are associated with elderly single person and retired couple households.

Table 8.12 shows the mean weighted sales by a household's level of community involvement. At the beginning of the panel study, in 1995, households with the middle-level of community involvement had more than twice as high a level of agricultural sales and households with the highest level of community involvement had almost four times the level of sales of the least community-involved households. This pattern also holds true

154 *Household Capital and the Agrarian Problem in Russia*

during the course of the panel study, even though the absolute level of community involvement for the total sample declined during that same period (see Chapter 5). These findings provide strong support for hypothesis 3b that households that are more integrated into village life will have higher levels of productivity.

Table 8.12: Mean Total Weighted Sales (kilograms) by Level of Community Involvement

Level of Community Involvement	1995	1996	1997
1-2.99 (low)	1,117 (n=83)	1,128 (n=75)	1,791 (n=88)
3.00-5.99 (middle)	2,308 (n=326)	2,242 (n=346)	3,912 (n=355)
6-7 (high)	4,296 (n=54)	5,142 (n=42)	6,732 (n=20)
Total (N=463)	2,326	2,325	3,630

1995: 17.47, $p<.001$. Scheffe, 1 x 2, $p<.01$, 1 x 3, $p<.001$, 2 x 3, $p<.001$; 1996:$F(2)=21.83$, $p<.001$; 1997:$F(2)=16.01$, $p<.001$. Scheffe 1 x 2 & 3, $p<.001$, 2 x 3, $p<.01$.

Table 8.13: Mean Total Sales (kilograms) by Number of Animals

Number of Animals	1995	1996	1997
0-50	128 (n=114)	234 (n=114)	291 (n=120)
51-325	1,470 (n=113)	1,154 (n=122)	2,037 (n=114)
326-611	2,811 (n=131)	2,677 (n=113)	4,160 (n=102)
612 +	5,030 (n=105)	5,319 (n=114)	7,791 (n=127)
Total (n=463)	2,326	5,319	3,630

1995:$F(3)=67.28$. 1 x 2, $p<.01$. Scheffe, 1 x 2, 3 & 4, $p<.001$, 2 x 3, $p<.01$, 2 x 3 & 4, $p<.001$, 3 x 4, $p<.001$; 1996:$F(3)=76.39$, $p<.001$. Scheffe, 1 x 3 & 4, $p<.001$, 2 x 3 & 4, $p<.001$, 3 x 4, $p<.001$; 1997:$F(3)=143.55$, $p<.001$. Scheffe, 1 x 2, 3 & 4, $p<.001$, 2 x 3 & 4, $p<.001$, 3 x 4, $p<.001$.

The level of household sales, like level of production, is most closely associated with the ability of the household to feed and care for livestock, (shown in Table 8.13). Earlier we showed that the level of production is strongly associated with the amount of livestock possessed by a household. Not surprisingly, Table 8.13 shows that this same strong relationship exists

Household Agricultural Production and Sales 155

between the number of animals in a household and its overall level of agricultural sales. In 1997, households with a weighted number of animals of 612 or more, for example, had almost 27 times more total agricultural sales than their fellow villagers who had a weighted number of animals of 50 or less. Households with the highest number of animals (612 or more) had almost twice as high sales as households with the next highest level of livestock (326-611). This provides strong support for hypothesis 7b.

Modeling the Effect of Household Capital on Sales

The effects of different types of household capital on overall household sales are shown in Figure 8.2. Obviously, the strongest direct effect on household sales is household production. Thus, all of the human and social capital variables that affect household production also, indirectly, affect household sales. This provides support for hypotheses 1b, 2b, and 3b. Moreover, Figure 8.2 also shows how the human and social capital variables operate indirectly through their association with physical capital to increase the advantages of households with more labor, a larger number of people in their networks, and those that are more involved in the activities of their communities. These paths support hypotheses 5b, 6b and 7b.

There are, however, three additional relationships shown in Figure 8.2 that could not be gleaned merely by looking at the relationships described earlier. First, community involvement, the indicator of integration into the larger village community, has a direct, and fairly strong effect on household sales. This is very interesting in light of the relationships shown in Figure 8.2 that identified those variables that affect household production. In particular, it appears that personal social networks (measured by the number of nonredundant ties) has a direct effect on production but not on sales. Alternatively, community involvement does not have a direct effect on production but it does have a strong direct effect on sales.

156 Household Capital and the Agrarian Problem in Russia

Figure 8.2: Combined Effects of Different Types of Household Capital on Household Agricultural Sales ($X^2=17.25$, $df=13$, $p=.19$, $GFI=.997$, $AGFI=.989$)

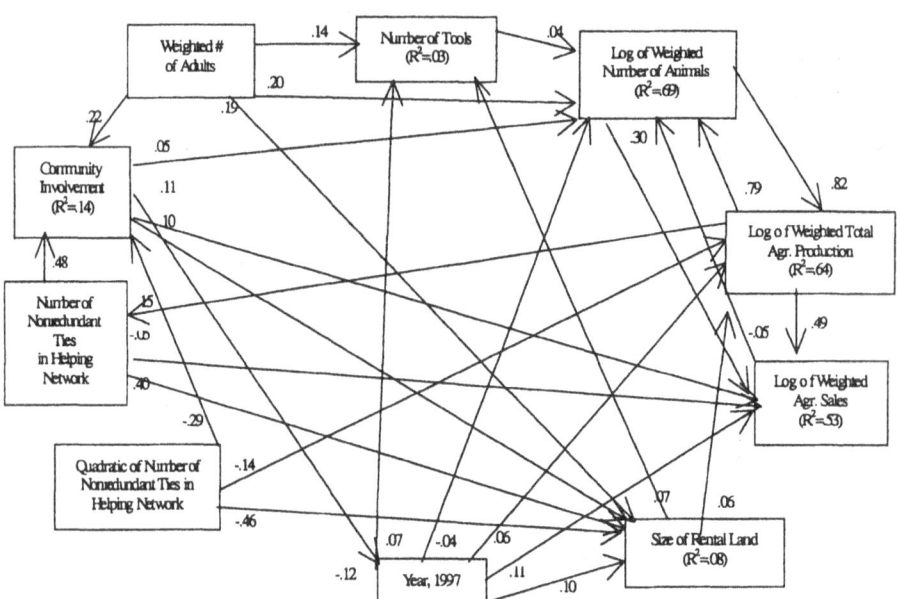

Our explanation for the apparently different effects or paths of the two purer forms of social capital is straightforward. High labor intensive production requires the strongest of strong ties, in the sense that persons who are feeding hogs or milking cows have to be very dependable day in and day out. This means that households with larger, but not too large, personal networks have an advantage in this regard.

Table 8.14: Standardized Regression Coefficients for Observed Exogenous Variables and Explained Variance (R^2) for Structural Model of Peasant Household Agricultural Production Sales

Observed endogenous	R^2	Standardized regression weights	Observed exogenous
Log of Weighted Agricultural Sales	0.53		
		0.49	Log of Weighted Total Agricultural Production
		0.30	Log of Weighted Number of Animals
		0.11	Community Involvment
		0.11	Year 1997
		-0.05	Number of Nonredundant Ties in Helping Network
Log of Weighted Agricultural Production	0.64		
		0.82	Log of Weighted Number of Animals
		0.15	Number of Nonredundant Ties in Helping Network
		-0.14	Quadratic of Number of Nonredundant Ties in Helping Network
		0.06	Size of Rental Land
		0.06	Year 1997
Log of Weighted Number of Animals	0.69		
		0.79	Log of Weighted Total Agricultural Production
		0.20	Weighted Number of Adults
		-0.04	Year 1997
		0.05	Community Involvement
		0.04	Number of Tools
Community Involvement	0.14		
		0.48	Number of Nonredundant Ties in Helping Network
		-0.29	Quadratic of Number of Nonredundant Ties in Helping Network
		0.22	Weighted Number of Adults
		-0.12	Year 1997
Size of Rental Land	0.08		
		0.40	Number of Nonredundant Ties in Helping Network
		-0.45	Quadratic of Number of Nonredundant Ties in Helping Network
		0.10	Community Involvement
		0.10	Year 1997
Number of Agricultural Tools	0.03		
		0.14	Weighted Number of Adults
		0.07	Size of Rental Land
		0.07	Year 1997

$X^2=17.25$, $df=13$, $p=.19$, $GFI=.997$, $AGFI=.989$.

On the other hand, sales of agricultural commodities require "weaker" ties in the sense that opportunities for sales may be obtained from persons with whom one has good relations, but these relations need not be too strong. Moreover, the kinds of activities required to bring products to market, such as asking a neighbor for a ride to the city or having a neighbor watch one's livestock for a day, do not require the same level of intensity of relationship that would be required for production. Thus, in this area households that have more expansive, but less intense relationships with a larger number of households have a distinct advantage in the emerging market economy. It is also important to note that households with higher levels of community involvement are also much more likely to rent land than are their neighbors.

All three forms of physical capital have indirect affects on household sales, and each of these types of physical capital are associated with one or more type of human and social capital. Thus there is support for the view that physical capital, in general, becomes a mechanism through which advantages or disadvantages in household human and social capital are multiplied. In this sense, the rich get richer and the poor get poorer, but wealth and poverty, in this instance, is based on access to human and social capital. These findings support hypotheses 5-7.

The most important type of physical capital, of course, is the amount of livestock in the household. In a peasant economy in which the sale of meat and dairy products is the primary way to raise cash, this is not surprising. The structural equation model shows clearly how increases in herds and flocks, which are derived from household human and social capital advantages, quickly produce a whole set of mutually reinforcing advantages for the household. Not only do households with more animals have higher levels of production of meat and milk, which, in turn means more sales, but also having more animals has a direct effect on sales. In this instance having more animals means that a household can afford to use a larger proportion of its meat for sale, whereas households not so advantaged must ration more of their meat and dairy products for household consumption. The negative arrow going from household sales to numbers of animals simply reflects the fact that slaughtering of animals for sale does reduce the size of a household's herds or flocks. The model shows that in 1997 there was an increase in household sales and a corresponding reduction in the number of animals per household.

Summary

The findings in this chapter provide support for the hypotheses in Chapter 2 that state that various types of human and social capital will differentiate between Russian peasant households in production and sales of agricultural commodities. Households with more household labor, larger numbers of persons in their social networks (up to a point), and higher levels of involvement in the festivals and ceremonies of the village and other families have a clear competitive advantage in the emerging market economy.

Moreover, the data show that advantages in human and social capital are multiplied by their association with a household's ability to gain access to certain types of physical capital, including mechanical agricultural equipment, rental of land, and ownership of livestock. The relationships between human and social capital, on the one hand, and physical capital, on the other, show that the economy of the contemporary Russian peasant household is not merely based on the type of hand labor that was typical in the nineteenth or early twentieth century. At the same time, however, more economically successful Russian peasant households, like smallholder households elsewhere, need to maintain the social capital (i.e, the social relationships that are based on moral bonds) that ensure contributions of individuals to collective household goals. Thus, even though household labor is operationally defined as human capital, it ultimately rests on social capital, i.e., household structure and relationships.

As we showed in Chapter 5, different quantities of household labor are embedded in different types of households. Households with the most stable and highest levels of labor are those that consist of husbands and wives living with their children and other adults.

Our most interesting findings are that the "purer" types of social capital, social networks and community involvement have a relatively independent impact on the production and sales of the households. This suggests that in spite of all of the conventional wisdom that Russians have become more isolated since the end of the "collectivist" Soviet period, more economically successful Russian peasant households have learned how to create social exchange helping networks and how to maintain their involvement in their village community.

Nevertheless, while social networks and community involvement may both fall under the general rubric of social capital (Coleman 1988; Wilson and Portes 1980), they arise from somewhat different sources and have somewhat different functions for households. The social networks of the households in the villages tend to be highly specialized and exclusive, at least to the extent that these networks are developed to obtain highly specific goals for a particular household. Not surprisingly, these networks tend to produce the most benefit for households in the highly arduous and time-consuming tasks related to the production of commodities.

Alternatively, involvement in the ceremonies and festivals of other families and the village as a whole tends to produce more inclusive, but, at the same time, less intense and less demanding social relationships. These kinds of relationships, which come closer to what American sociologists would refer to as "weak ties" (Granovetter 1973) have been declining among the sample as a whole, from 1995 to 1997, but among the more economically successful have remained in tact.

In later chapters we will look at how these different types of social capital are impacting on the mental health and quality of life of Russian peasants and how they might eventually affect the development of a true civic culture in rural Russia. At this point, however, let us turn to an examination of all of the sources of income, both monetary and nonmonetary, that Russian peasant households currently receive.

9 Employment, Household Income and Durable Goods

David J. O'Brien, Valeri V. Patsiorkovski and Larry D. Dershem

Introduction

In the previous chapter, we showed how different types of household capital affect a household's ability to produce and sell agricultural commodities. Agricultural sales, along with income generated from other types of family business enterprises, such as sales of services or goods, are a significant portion of total household income for most rural Russian households. Thus, we would expect to find that households with higher amounts of human and social capital would also have higher overall household incomes.

In order to get a complete picture of household income differentiation, however, it is necessary to examine two other sources of rural household income, salaries and transfer payments. Salaries are obtained from employment, which in the Russian village traditionally has meant work in the large agricultural enterprises, the *kolkhozy* or *TOO*s. Transfer payments include small allowances for pregnant women, for children from birth until age eighteen (as long as they are enrolled in school), stipends for children who attend college, and alimony. This income is a relatively small portion of the total amount of transfer payments made to households in the three villages. In Bolshoe Sviattsovo, for example, only one child was born in 1996. Moreover, the central government in Moscow has become increasingly less willing to make these payments and thus they may eventually be eliminated.

The main source of transfer payment income in the Russian village is pensions for retired elderly persons (for women 55 years of age and older, for men 60 years of age and older). These transfer payments are roughly eight to nine times greater than those provided to pregnant women and children. Moreover, from one-quarter to one-third of the households in the

162 *Household Capital and the Agrarian Problem in Russia*

sample depends on this type of transfer payment as their primary source of income.

Employment Patterns from 1995 to 1997

There were some modest, but nonetheless important, shifts in the distribution of different types of employment statuses in the three villages between 1995 and 1997, which are shown in Table 9.1. This table shows the percentage of adults and gender composition (in percentage of males) in each employment status.

Table 9.1: Employment Status of Working Age Adults in Three Russian Villages in 1995 and 1997

Employment Status	1995 (N=959)		1997 (N=833)	
	Percentage of Adults in Status	Percentage Males	Percentage of Adults in Status	Percentage Males
Full-time	51.1	58.6	62.3	53.4
Part-time	4.4	33.3	2.8	52.4
Unemployed	1.7	56.3	2.8	66.7
Retired	34.8	24.6	25.9	41.8
Handicapped	1.9	66.7	4.1	52.9
Homemaker	2.5	8.3	0.4	33.3
Childcare leave	2.5	0.0	0.7	0.0
Student	1.1	54.5	1.1	55.6

The percentage of adults employed full-time increased by slightly more than 11 percent during the three-year period. This is coupled with a corresponding decline in the number of persons listed as working part-time, retired, homemakers, or on childcare leave. This is an adaptation of households to increased pressures to gain income from any available sources. In part, this adjustment appears to have been a case of older women leaving retirement and younger women leaving homemaking or ending childcare leave and becoming employed full-time. The percentage of full-time workers who are males decreased by slightly more than five percent during the three-year period.

In 1995, at the beginning of the panel study, 95 percent of the working-age adults in the households in the sample worked in their village. This percentage remained almost unchanged during the three-year period of the study, dropping only eight-tenths of a point to 94.2 percent from 1995 to 1997. There was, however, a slight increase in the number of persons who commuted to work in a city. This number rose from ten in 1995 to eighteen in 1997. Given the difficulties in transportation and the relative isolation of these villages, it is unlikely that the proportion of commuters will increase very much in the near future. For all practical purposes these villages are relatively self-contained rural communities. This is quite different than the typical pattern in American communities, for example, where the vast majority of the working-age rural population would be working outside of their local village.

Table 9.2: Types of Enterprises Where Working-Age Adults Are Employed in Three Russian Villages from 1995 to 1997

Types of Enterprises	1995 (N=513)	1996 (N=513)	1997 (N=504)
Large Enterprise *(kolkhoz or TOO)*	80.3	73.1	70.4
Public Services	13.4	20.1	21.4
Fermer (officially registered private farmer)	2.7	2.1	3.2
Household Enterprise	3.5	4.7	5.0

Table 9.2 shows changes in the types of enterprises and organizations in which persons worked in the three villages during the three-year panel study. From 1995 to 1997, almost ten percent of the labor force moved from employment in the large enterprises to some other type of employment. The largest increase in employment is in the category of "public services," which includes nurses, teachers, local government workers and social workers. Most of the change in this category is due to the federal government's program to place social workers in the villages. For persons over 80 years of age, without family members in the village, these services are provided free of charge. Other persons who need assistance, but have relatives in the village, are required to pay fees for these services.

There is a slight increase in the number of officially registered *fermery* from 1995 to 1997, which is different than the slight decline found in the country as a whole (see Chapter 2). The most interesting change occurred in the number of employed adults who listed the type of organization in which they worked as some type of household enterprise. This includes agricultural businesses, such as a household processing plant for producing sunflower oil from seed, service businesses, such as clothing, household appliance or vehicle repair, or shops for selling food, dry goods or hardware supplies.

The vast majority of working-age adults still work for the large enterprises. This reflects the fact that the large enterprises still have not trimmed their work forces to become more efficient. Moreover, the situation where the majority of working-age adults are employed in farming is at odds with typical employment patterns in American or Western European rural villages. In these other national contexts, most rural residents work outside farming. Nevertheless, it is significant that employment patterns are changing in the Russian villages, albeit slowly, and our expectation is that they will become closer to rural employment patterns in the United States and Western Europe as time goes on.

It is important to recall that during this same period, employment in the large enterprises declined by over ten percent. Thus, it would appear that a significant portion of the newly employed persons, especially women, were being employed in organizations other than the large enterprises, including public administration and household enterprises (see Table).

The significance of the shifts in employment *within the villages* cannot be overemphasized. The fact that there is almost no commuting to urban centers for work and that there are no industries moving into the villages from the outside makes even more impressive the ten percent shift in employment from the large enterprises to other organizational settings. This change highlights the institutional restructuring in Russian villages that was shown earlier in Figure 1.2.

Sources of Household Income

The different sources of income for the households in our sample are shown in Table 9.3. The figures in the columns on the left are calculated on the basis of monetized income only. The columns on the right are based on calculations that include nonmonetized income, in the form of agricultural products that are produced *and consumed* by the households (Rose and McAllister 1996). The proportion of total income generated by primary salaries, which is gained, by and large, from working for the large enterprises or local government, did not change much from 1995 to 1997. It was virtually the same if we consider only monetized income, but increased slightly if we consider both monetized and nonmonetized income. There was, however, an increase in secondary salary. In Latonovo, for example, a number of persons who work for the *TOO* have an additional part-time job working for either poultry or milk processing facilities in a nearby village.

Table 9.3: The Contribution of Different Sources of Income (Monetized and Nonmonetized) to Total Monthly Household Income (Mean Number of Rubles, Adjusted for Inflation) in Three Russian Villages from 1995 to 1997 (in percent)

		1995		1996		1997	
Salary & Wages	Primary salary	30.6	19.0	35.1	21.9	30.5	21.0
	Secondary salary	1.0	1.0	4.6	2.9	4.2	2.9
Transfer payments		33.2	20.2	34.9	21.7	27.6	18.9
Household Enterprises	Business	6.0	3.7	4.5	2.9	9.1	6.3
	Benefits	3.2	1.9	6.5	4.1	4.1	2.8
	Agricultural sales	27.3	16.6	19.0	11.8	28.5	19.5
Nonmonetized consumption		----	39.2	----	37.7	----	31.6
Mean (N=463)		407,247	669,656	337,804	541,934	350,412	512,077

Employment, Household Income and Durable Goods 167

The proportion of income derived from transfer payments decreased from 1995 to 1997. In part, this drop reflects the limited capacity of the Russian central government to maintain transfer payments that keep up with inflation. But, this is not the total explanation. During the three year-period of the panel study, households that had depended almost exclusively on retirement pensions have learned new ways to generate income from other sources. Among these households, the proportion of income derived from some type of household enterprise was 19 percent of their total income in 1995 but rose to 31 percent of their income in 1997. This is due, in part, to an increase in agricultural sales, which accounted for 15.6 percent of income for this type of household in 1995 and 20 percent of their income in 1997. But there also were increases in benefits, accounting for 2.6 percent of household income in 1995 and 5.2 percent of household income in 1997. This is the result of an increase in the amount of informal rental agreements in which pensioners have allowed neighbors to cultivate their household plots in return for some share of the harvest. Finally, among this retired group there was an increase in the proportion of household income from non-agricultural businesses; from 1.0 percent in 1995 to 5.7 percent in 1997.

All types of household enterprises (other benefits, non-agricultural benefits and agricultural sales) contributed slightly more than one-third (36.5 percent) of all monetary income in peasant households in 1995, as shown in Table 9.3. If we consider goods and services that are produced by households for both *sale and consumption*, i.e., monetized and nonmonetized income, then household enterprises contributed almost two-thirds (61.4 percent) of total household income in that year. This proportion declined in 1996, due to specific problems that occurred in the village of Latonovo.

For several years the *TOO* in Latonovo had been assisting peasant households in the collection, storage and transportation of milk and eggs they had produced. In effect, the *TOO* had been a broker that provided the resources to bring peasant household production of milk and eggs to market on a daily basis, which was practically impossible for the vast majority of households to do on their own. This provided households with a vital source of daily cash income. In 1996, however, the chairman of the *TOO* decided that he would no longer assist peasant households in the way just described. This produced a dramatic drop in sales of these commodities in that village and reduced the overall level of income from sales for the total sample in that year. Eventually, the households in Latonovo were successful in staging a rebellion against the chairman of the *TOO*, forcing his resignation and subsequent replacement by a man much more sympathetic to household production and sales. There was a subsequent increase in peasant household income from the sale of milk and eggs in 1997 in Latonovo.

Despite the obvious and expected fluctuations in household enterprise income, our view is that the most important overall trend in income in the Russian villages is for household enterprises to contribute a larger share of a household's monetary income. Considering only monetized income sources, the contribution of household enterprises rose from 36.5 to 41.7 percent in just three years.

Table 9.4 shows the distribution of different types of household income by village. The average salary and wages in these villages is one-fourth the average salary and wages in Russia as a whole, which were 472,392 in 1995, 835,924 in 1996, and 1,200,000 in 1997 (Russia in Figures 1997: 9; Rossia 1997:138). In fact, as shown in Graph 9.1, the accrued wages of agricultural workers (366,862) in 1996 was merely 23.2 percent of workers in finance (1,582,091), 32.9 percent of workers in industry (1,115,733) and 67.8 percent of workers in education (Russia in Figures 1997:Table 4.11, pg. 59). Moreover, much of the salary and wages, shown in Table 9.4 were paid after delays of many months.

Graph 9.1: Comparison of Accrued Wages for Russian Workers by Selected Sectors from 1991 to 1996

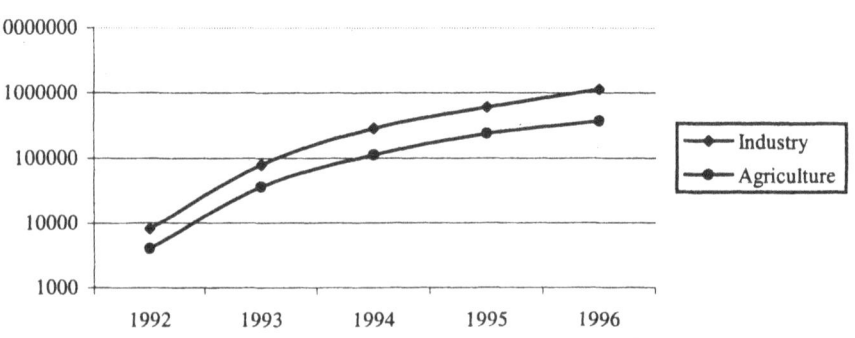

Source: *Russia in Figures* 1997:Table 4.11, pg. 59.

Table 9.4: Income from Different Sources by Village in Three Russian Villages from 1995 to 1997 (Mean Number of Rubles, Adjusted for Inflation)

	Salary & Wages			Transfer Payments			Household Enterprises		
	1995	1996	1997	1995	1996	1997	1995	1996	1997
Latonovo (n=157)	85,503	92,127	94,796	122,751	105,200	95,141	155,458	125,616	137,900
Vengerovka (n=156)	155,756	130,897	106,608	118,551	118,527	97,001	155,949	106,378	172,957
Sviattsovo (n=150)	130,562	133,929	120,442	165,140	130,270	98,353	133,180	70,441	128,043
Total (N=463)	123,772	118,733	107,084	135,069	117,813	96,809	148,406	101,259	146,519

Salary & Wages by Village, 1995 F(2)=7.78, p<.001; Scheffe Latonovo by Vengerovka, p<.001, Latonovo by Sviattsovo Bolshoe, p<.05. 1996 F(2)=4.07, Scheffe, Latonovo by Sviattsovo Bolshoe, p<.05. 1997 F(2)=1.29, p=ns.
Transfer Payments by Village, F(2)= 1995 F(2)=7.25, p=.001, Scheffe, Latonovo x Sviattsovo Bolshoe, p<.01. 1996 F(2)=2.99, p=.05; Scheffe Latonovo by Sviattsovo Bolshoe p<.05. 1997 F(2)=0.8, p=ns.
Household Enterprise by Village: 1995 F(2)=.36, p=ns; 1996 F(2)=6.19, p =.01; Scheffe Latonovo by Sviattsovo Bolshoe, p<.04. 1997 F(2)=3.04, p<.05, Scheffe, Vengerovka by Bolshoe Sviattsovo, p=.07 .

Income from salaries and wages is significantly higher in Vengerovka and Bolshoe Sviattsovo than in Latonovo in 1995 and significantly higher in Bolshoe Sviattsovo than in Latonovo in 1996. By 1997, however, there is no statistically significant difference between villages in income received from salary and wages. There is an increase in the mean level of salary and wages in Latonovo and a decrease in that type of income in Vengerovka and Bolshoe Sviattsovo. The differences in average salary and wages between the three villages are not associated with the economic viability of their respective large agricultural enterprises. All three large agricultural enterprises in these three villages have substantial unpaid debts. Moreover, the large agricultural enterprise in Vengerovka went bankrupt, and reorganized, in 1997. The differences in salary and wages are associated with two other issues. First, with the type of commodity produced by the agricultural enterprise. The enterprise in Latonovo specializes in grain and sunflower, while the other enterprises are more diversified. In addition to producing grain, the latter produce milk, meat, buckwheat and hay enabling them to supply local niche markets. Second, there is a regional difference in the cost of living, which is lower in southern Russia where Latonovo is located.

The total amount of income received from transfer payments decreased substantially (39.5 percent) in all three villages from 1995 to 1997. On average, households in Latonovo had higher levels of transfer payment income than did households in Bolshoe Sviattsovo in 1995 and in 1996, but there were no significant differences between villages on the amount of income received from transfer payments in 1997. The larger average transfer payment in Sviattsovo is due to a different demographic structure of households (i.e., a greater number of elderly people) than in the other two villages.

Income from household enterprises fluctuated considerably during the panel study. In 1995 there was no statistically significant difference between villages in this type of income. During the three-year period, however, there was some evidence of a growing differentiation between the three villages in income gained from household enterprises. From 1995 to 1996, income from household enterprises decreased in all three villages, but the reduction was most severe in Bolshoe Sviattsovo. In 1996, Bolshoe Sviattsovo had lower average levels of income from household enterprises than the other two villages, although the only difference that reaches the $p<.05$ level of statistical significance is between this village and Latonovo.

The largest gain in income from household enterprises is in Vengerovka, almost a two-thirds (62.6 percent) increase from 1996 to 1997. This is a result of households being involved in a new type of credit program.

The Belgorod provincial government has instituted a special credit program to assist peasant households to improve existing homes or to build new homes and buildings for storing grain, silage, or for keeping animals. The fund, which is called, *"fond podderrzhki individualnogo zhilishhchnogo stroitelstva na sele"* (fund for the support of individual buildings in rural areas) lends money to peasant households and they repay their debt in agricultural products. These products include meat, milk, eggs, cottage cheese, or sour cream, which can be produced by the peasant household. The results obtained in Vengerovka are mirrored by the increases in production by households in all villages in the province that participated in this program. In Belgorod *Oblast* meat produced by households increased from 48 percent in 1996 to 61 percent in 1997 and milk increased from 34 percent to 41 percent during this same time period. Households that participated in this program received the equivalent of 25 million USD in 1997 (*Agronpromyshlennyi kompleks Belgorodskoi oblasti sostoianie i perspektivy* 1998).

Income and Household Capital

Our primary focus in this section is to estimate how different types of household capital affect various sources of income, and in turn, total household income. First, we will consider household labor and sources of household income (see Table 9.5). Second, we will consider the two "purer" forms of social capital, community involvement (see Table 9.6) and personal helping networks (see Table 9.7) and their relationship to sources of income and total household income.

Table 9.5 shows the distribution of different types of income in households with different levels of household labor. Overall, there is a decline in the amount of income derived from salary and wages and from transfer payments, but income from household enterprises remains fairly stable. Even those households with the lowest amount of household labor (scores from 0 to 1.74) managed to increase the amount of money they made from household enterprises during the three-year period. For elderly widows or retired couples this meant earning income from rental of the land from their household plot. These adaptations are evidence of the enormous resilience of Russian peasant households and their ability to cope with difficult economic circumstances.

Nonetheless, households with more available working age adult members are clearly at an advantage over their neighbors. The quantity of household labor is positively associated with income from salary and wages and household enterprises, but negatively associated with transfer payments. This is consistent with what was shown in Chapter 8. In the case of both salary and wages and household enterprise income, the most dramatic gains are observed from the first to the second levels of household labor. This illustrates how close to one-third of the village households (31.7 percent) consists of single adults or elderly retired couples who have very limited income from salary and wages or from household enterprises. Income from household enterprises, on average, contributes almost two-thirds of the total household income in the Russian villages (see Table 9.3). This places economically vulnerable groups, such as the elderly in a particularly disadvantaged position vis-à-vis their neighbors.

Community involvement, participation in other families and village events, also has an affect on sources of income and total household income.

Table 9.6 shows the relationships between different levels of community involvement and the amount of different types of income received by a household. Higher levels of community involvement are positively associated with higher salary and wages and negatively associated with higher levels of transfer payments. In part, this is due to the fact that households with higher levels of community involvement tend to be comprised of younger members, while households with low levels of community involvement are likely to be comprised of elderly single persons and retired couples. As we will see later, however, even when the composition of the household is controlled for (i.e., level of household labor) level of community involvement is positively associated with higher levels of income from salary and wages.

Employment, Household Income and Durable Goods 173

Table 9.5: Income from Different Sources by Household Labor in Three Russian Villages from 1995 to 1997 (Mean Number of Rubles, Adjusted for Inflation)

	Salary & Wages			Transfer Payments			Household Enterprises		
	1995	1996	1997	1995	1996	1997	1995	1996	1997
0-1.74	13,175 (n=143)	14,603 (n=146)	8,896 (n=147)	170,732 (n=143)	143,957 (n=146)	112,959 (n=147)	47,790 (n=143)	43,167 (n=146)	53,052 (n=147)
1.75-2.74	129,451 (n=164)	127,686 (n=177)	119,200 (n=173)	138,113 (n=164)	108,164 (n=177)	86,280 (n=173)	187,531 (n=164)	107,303 (n=177)	167,295 (n=173)
2.75-3.74	200,356 (n=108)	220,892 (n=95)	182,906 (n=93)	79,096 (n=108)	81,734 (n=95)	77,326 (n=93)	190,222 (n=108)	137,886 (n=95)	211,240 (n=93)
3.75-4.74	252,882 (n=34)	195,135 (n=35)	209,909 (n=44)	126,980 (n=34)	135,838 (n=35)	121,233 (n=44)	204,824 (n=34)	165,196 (n=35)	210,498 (n=44)
4.75+	282,571 (n=14)	242,629 (n=10)	234,097 (n=6)	186,571 (n=14)	186,534 (n=10)	127,534 (n=6)	258,214 (n=14)	270,677 (n=10)	365,041 (n=6)
Total N=463	123,772	118,733	107,085	135,069	117,813	96,809	148,406	101,259	146,519

Salary & Wages by Human Capital Groups: 1995 $F(4)=42.96$, $p<.001$; Scheffe 1 by 2, $p<.001$; 1 by 3, $p<.001$; 1 by 4, $p<.001$; 1 by 5, $p<.001$; 2 by 3, $p<.01$, 2 by 4, $p<.001$, 2 by 5, $p<.01$. 1996 $F(4)<49.79$, $p<.001$; Scheffe 1 by 2, $p<.001$, 1 by 3, $p<.001$, 1 by 4, $p<.001$, 1 by 5, $p<.001$, 2 by 3, $p<.01$, 2 by 4, $p<.001$. 1997 $F(4)=44.66$, $p<.001$; Scheffe 1 by 2, $p<.001$, 1 by 3, $p<.001$, 1 by 4, $p<.001$, 1 by 5, $p<.001$, 2 by 3, $p<.01$, 2 by 4, $p<.001$.
Transfer Payments by Human Capital Groups: 1995 $F(4)=10.62$, $p<.001$, Scheffe 1 by 2, $p<.01$, 1 by 3, $p<.001$, 3 by 4, $p<.05$, 3 by 5, $p<.01$. 1996 $F(4)=9.86$, $p<.001$; Scheffe 1 by 2, $p<.01$, 1 by 3, $p<.001$, 3 by 4, $p<.05$, 3 by 5, $p<.01$. 1997 $F(4)=6.57$, $p<.001$; Scheffe 1 by 2, $p<.05$, 1 by 3, $p<.01$.
Household Enterprise by Human Capital Groups: 1995 $F(4)=8.11$, $p<.001$, Scheffe 1 by 2, $p<.001$, 1 by 3, $p<.001$, 1 by 4, $p<.05$. 1996 $F(4)=9.86$, $p<.001$; Scheffe 1 by 2, $p<.001$, 1 by 3, $p<.01$, 1 by 4, $p<.001$, 3 by 4, $p<.05$, 3 by 5, $p<.01$. 1997 $F(4)=23.14$, $p<.001$; Scheffe 1 by 2, $p<.001$, 1 by 3, $p<.001$, 1 by 4, $p<.001$, 1 by 5, $p<.001$, 2 by 5, $p<.05$.

Table 9.6: Income from Different Sources by Level of Community Involvement in Three Russian Villages from 1995 to 1997 (Mean Number of Rubles, Adjusted for Inflation)

	Salary & Wages			Transfer Payments			Household Enterprises		
	1995	1996	1997	1995	1996	1997	1995	1996	1997
1.00-2.99 (low)	72,265 (n=83)	70,219 (n=75)	43,900 (n=88)	171,747 (n=83)	144,439 (n=75)	113,751 (n=88)	74,313 (n=83)	59,2442 (n=75)	84,978 (n=88)
3.00-5.99 (middle)	120,520 (n=326)	120,836 (n=346)	119,592 (n=355)	132,807 (n=83)	113,554 (n=346)	94,067 (n=355)	145,660 (n=326)	101,757 (n=346)	156,435 (n=355)
6.00-7.00 (high)	222,574 (n=54)	188,034 (n=42)	163,083 (n=20)	92,352 (n=83)	105,351 (n=42)	70,928 (n=20)	278,870 (n=54)	172,183 (n=42)	241,275 (n=20)
Total N=463	123,772	118,737	107,085	135,069	117,813	96,809	148,406	101,259	146,519

Salary & Wages by Community Involvement, 1995 F(2)=15.18, p=.001; Scheffe Low by Middle, p<.05, Low by High, p<.001, Middle by High p<.001. 1996 F(2)=9.35, p<.001; Scheffe Low by Middle, p<.05, Low by High, p<.001, Middle by High, p<.05. 1997 F(2)=12.59, p<.001; Low by Middle, p<.001, Low by High, p<.01.
Transfer Payments by Community Involvement, F(2)= 1995 F(2)=7.62, p<.001, Scheffe Low by Middle, p<.03, Low by High, p<.001. 1996 F(2)=4.01, p<.05; Scheffe Low by Middle, p<.05. 1997 F(2)=4.21, p<.05. Low by High, p<.05.
Household Enterprise by Community Involvement, 1995 F(2)=10.10, p=.001; Scheffe Low by High, p<.001, Middle by High, p<.01; 1996 F(2)=8.98, p<.001; Scheffe Low by Middle, p<.05, Low by High, p<.001, Middle by High, p<.05; 1997 F(2)=9.98, p<.001, Scheffe Low by Middle, p<.001, Low by High, p<.002.

Employment, Household Income and Durable Goods 175

The most important effect of community involvement is to increase income from household enterprises. Community involvement certainly has a socio-emotional benefit. Households that participate in the life of their village community receive important non-material benefits that improve their quality of life. In the next chapter we will show that persons who feel that they "fit into" their village community have lower levels of symptoms of stress than do their neighbors. But, community involvement also has a very practical utilitarian function as well. Households that have a middle-level of involvement in the festivals and ceremonies of other families and the village as a whole, on average, receive almost twice as much income from household enterprises as do their neighbors with the lowest level of community involvement. Families that are involved in community life are more likely than other families to know about opportunities for processing or selling products in markets outside their village. These families also are much more likely to work with neighbors in informal marketing or processing arrangements that increase their household income.

The critical social capital function of community involvement is to link the household to opportunities and resources outside of itself. As shown in Chapter 8, even though agricultural production is primarily affected by the level of labor in the household, the ability of the household to sell what it produces is dependent to a significant degree on its extra-household linkages, among the most important of which is involvement in different social activities such family and village ceremonies and events.

Similar to community involvement, the size of nonredundant helping networks has a direct effect on sources of income and total household income.

Table 9.7: Income from Different Sources by Size of Non-Redundant Helping Networks in Three Russian Villages from 1995 to 1997 (Mean Number of Rubles, Adjusted for Inflation)

	Salary & Wages			Transfer Payments			Household Enterprises		
	1995	1996	1997	1995	1996	1997	1995	1996	1997
0-3	79,559 (n=102)	63,925 (n=91)	75,169 (n=89)	159,441 (n=102)	128,627 (n=91)	97,020 (n=89)	83,147 (n=102)	54,804 (n=91)	95,327 (n=89)
4-5	128,213 (n=155)	140,701 (n=145)	97,426 (n=133)	128,263 (n=155)	117,649 (n=145)	101,921 (n=133)	130,387 (n=155)	88,091 (n=145)	119,694 (n=133)
6	154,218 (n=66)	128,018 (n=74)	106,107 (n=79)	143,015 (n=66)	115,761 (n=74)	98,558 (n=79)	164,409 (n=66)	117,264 (n=74)	191,776 (n=79)
7-10	136,984 (n=122)	126,516 (n=139)	136,001 (n=144)	123,257 (n=122)	111,738 (n=139)	89,304 (n=144)	203,951 (n=122)	139,239 (n=139)	179,243 (n=144)
11-17	134,889 (n=18)	121,104 (n=14)	109,215 (n=18)	106,500 (n=18)	120,337 (n=14)	110,351 (n=18)	238,222 (n=18)	77,908 (n=14)	137,243 (n=18)
Total N=463	123,772	118,733	107,084	135,069	117,813	96,809	148,406	101,259	146,519

Salary & Wages by Nonredundant Helping Network Groups: 1995 F(4)=2.78, p=.05: Scheffe 1 x 3, p=.08, 1996 F(4)=4.38, p=.01: Scheffe 1 x 2, p=.01, 1 x 4, p=.05; 1997 F(4)=2.90, p=.05: Scheffe 1 x 4, p=.05, 2 x 3, p=ns, 2 x 4, p=ns, 3 x 4, p=ns
Transfer Payments by Nonredundant Helping Network Groups: 1995 F(4)=1.82, p=ns. 1996 F(4)=0.49, p=ns: 1997 F(4)=0.76, p=ns.
Household Enterprise by Nonredundant Helping Network Groups: 1995 F(4)=3.70, p=.01: Scheffe 1 x 4, p=.05; 1996 F(4)=5.90, p=000; Scheffe 1 x 4, p=.000, 2 x 4, p=.05; 1997 F(4)=5.91, p=.000; Scheffe 1 x 3, p=.01, 1 x 4, p=.01, 2 x 3, p=.05

The relationships between household helping networks, measured in the number of nonredundant ties, and different sources of income are shown in Table 9.7. There is a curvilinear relationship between the size of a household's helping network and its income from salary and wages. Increases in the size of helping networks up to 6 in 1995 and 10 in 1997 are associated with higher income levels. Increases in network size after six in 1995 and ten in 1997 are associated with declines in income from salary and wages. Extremely large nonredundant tie networks are associated with households consisting of elderly dependent persons (see Chapters 6 and 8). There is no relationship between nonredundant ties and transfer payments. This indicates that there is considerable variation in the network structure of older persons who are the most likely to receive pensions.

In 1997 there is a curvilinear relationship between the size of a household's helping network and the income it receives from household enterprises. The statistical nature of this relationship is similar to that observed for salary and wages, except that the point in which additional helping network members begins to bring a reduction in income is earlier, after six persons, than in the case of salary and wages. *One of the most important substantive points to be noted here is that the strength of association between the size of a household's helping network and income derived from household enterprises is weaker than the association between community involvement and income derived from household agricultural enterprises.* The second substantive point is that these different types of household capital play different roles in household production and sales. *Personal networks are more important in production whereas community involvement is more important in sales* (see Chapter 8). The relative impact of these different types of household capital on household income will be discussed in the next section.

Modeling the Effect of Household Capital on Income

A structural equation model was utilized to understand the complex interrelationships between household capital, sources of income, and total household income. Figure 9.1 shows the results of the AMOS structural equation model of the effects of different types of household capital on total household *monetary* income, through the various components of that income. The model provides strong support for hypothesis 8 (see Chapter

1) that households with higher levels of household labor, community involvement and personal helping networks will have higher levels of income.

Because of the complexity of the income model, indicators of household physical capital are not included. As can be shown in Chapters 7 and 8, however, levels of physical capital are strongly associated with the other components of household capital, household labor, community involvement and nonredundant social network ties.

The standardized regression weights show that the proportion of total household income contributed by salary and wages and household enterprises is virtually equal. All three of the household capital variables included in the model have a direct effect on total household income. The strongest effects come from the size of personal helping networks and the weighted-number of adults in the household. The positive sign on the size of personal networks reflects the positive contribution of increases in network size up to nine persons. Those households that fall at the higher end of this continuum, seven to nine nonredundant ties, typically are made up of younger energetic couples with children. The negative sign on the quadratic term for size of helping network indicates (see Chapters 6 and 8) that households with extremely high numbers of persons in their helping networks are most likely to be made up of more dependent persons, especially elderly widows. The direct effect of community involvement on total household income is weaker than the effect of the other forms of household social capital. It is nonetheless important to recognize that households that expand their community connections appear to do better economically, even when we control for the effects of household labor and personal social networks. This means that there are some positive economic incentives for households to contribute to community institution building and maintenance in a way that may ultimately assist in the development of a civic culture in rural Russia. In short, the empirical data show that economic success does not require households to become more self-centered in their outlook.

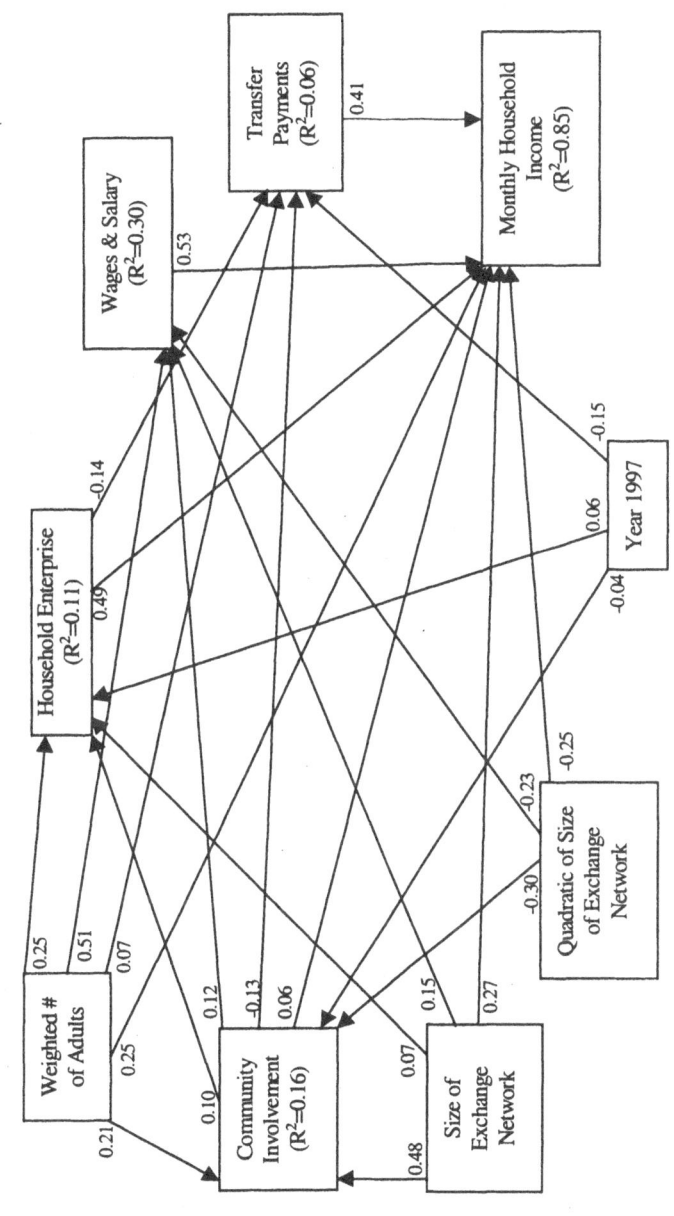

Figure 9.1: Effects of Household Labor, Helping Networks and Community Attachment on Peasant Household Monetary Income from 1995-1997 ($X^2=7.45$, $df=9$, $p=0.59$, Goodness of Fit=0.99, Adjusted Goodness of Fit=0.99)

180 Household Capital and the Agrarian Problem in Russia

Table 9.8: Standardized Regression Coefficients for Observed Exogenous Variables and Explained Variance (R^2) for Structural Model of Peasant Household Monetary Income

Observed endogenous	R^2	Standardized Regression weights	Observed exogenous
Monthly household income	0.85		
		0.53	Salary & wages
		0.49	Household enterprise
		0.41	Transfer payments
		0.27	Size of helping network
		-0.25	Quadratic of size of helping network
		0.23	Weighted number of adults
		0.06	Community involvement
Salary & wages	0.30		
		0.51	Weighted number of adults
		0.15	Size of helping network
		-0.23	Quadratic of size of helping network
		0.12	Community involvement
Transfer payments	0.06		
		-0.15	Year 1997 compared to 1995
		-0.14	Household enterprise
		-0.07	Weighted number of adults
Household enterprises	0.11		
		0.25	Weighted number of adults
		0.10	Community involvement
		0.07	Size of helping network
		0.06	Year 1997 compared to 1995
Community involvement	0.16		
		0.48	Size of helping network
		0.21	Weighted number of adults
		-0.14	Year 1997 compared to 1995
		-0.13	Transfer payments

$X^2=7.45$, df=9, p=0.59, *GFI*=0.99, *AGFI*=0.99.

As would be expected, weighted number of adults is strongly associated with the level of salary and wages received by the household. It can be observed, however, that the two purer forms of social capital, size of helping network and community involvement have weaker, but nonetheless significant effects on the extent to which a household can obtain income from salary and wages.

The three variables having a direct effect on transfer payments are predictable. The level of government support for transfer payments has not kept up with inflation and thus the regression coefficient shows that there is a decline from 1995 to 1997. Household enterprises, which are most prominent in younger households, are negatively associated with transfer payments, as is the weighted number of adults in the household.

All three types of household capital are positively associated with income from household enterprises. This provides a more general confirmation of the findings on a specific type of household enterprise, agricultural sales, that was shown in Chapter 7.

Finally, the model shows some intriguing paths to level of community involvement. As shown earlier, the overall level of community involvement has declined during the course of the panel study. This is shown in the negative path from year 1997 to community involvement. Nonetheless, households with high human labor (i.e. weighted number of adults) and larger size helping networks are more involved in the institutional life of their village community. Again, this is very encouraging with respect to the development of civic culture in Russian villages.

Income Differentiation and Poverty

A steep rise in poverty and income differentiation has accompanied the transition to a market economy in Russia. Many scholars and research institutes have studied the scale and profile of poverty in Russia in the post-Soviet period (Klugman 1997).

According to official Russian government data, the coefficient of differentiation in income was 5.5 in 1991, increasing to 11 in 1993 and 13 in 1994 (*Rossia* 1997:140). Countrywide data, which is based largely on urban areas where 73 percent of the population lives, shows that the top ten percent of incomes received 13 times more income than the lowest ten percent in 1994. There was a slight decrease in income differentiation from 1995 to 1997, estimated to be between 10 to 13 (*Rossia* 1997: 143).

Table 9.9 presents the coefficient of differentiation of income for the households in the three villages in our panel study.

Table 9.9: Differentiation of Income in the Three Villages by Year

	1995	1996	1997
Mean income for bottom ten percent income group	174,474	170,773	146,673
Mean income for the top ten percent income group	1,636,315	1,100,599	825,114
Coefficient of Differentiation*	9.4	6.4	5.6

*Coefficient of differentiation is calculated by dividing the mean income of the 10th decile income group by the mean income of the 1st decile income group.

Income differentiation in these rural villages is lower than in the country as a whole. In the three villages, the differentiation coefficient was 9.4 in 1995, decreasing to 5.6 by 1997. The decrease in income differentiation among households over the three years is primarily the consequence of two factors. First, in rural areas, there are few alternative means to rapidly increase household income. Second, the main way to increase household income, selling household agricultural production, can be accomplished by most households.

Measured by the minimum consumption basket standard, it was estimated that approximately 70 to 80 percent of all Russian families lived in poverty in 1995 (Rimashevskaya 1997:121). Table 9.10 shows the per capita minimum consumption basket, the percentage of households below the minimum using monetized income, and the percentage of households below the minimum when combining monetized and nonmonetized sources of income for households in the three rural villages from 1995 to 1997. The proportion of households in the three villages that were below the official minimum poverty line during this time period was higher than the national average when based only on monetized income. This is in keeping with the lower accrued monthly wages of agricultural workers compared to workers in other sectors (see Graph 9.1 above).

Table 9.10: Percentage of Households in Poverty in the Panel Study

	1995	1996	1997
Per capita minimum consumption basket*	264,100	369,400	411,000
% households living in poverty by monetized income	91.1	83.8	69.8
% households living in poverty by monetized & nonmonetized income	62.4	42.3	28.9

* Source: *Rossiia* 1997:119.

The trend, however, is for the percentage of households living below the poverty line to decrease, by approximately 20 percent during these three years. This decline was not the result of increases in salary and wages, but from increases in income from household enterprises (see Table 9.3).

When including nonmonetized sources of income to total household income, the percentage of households living below the poverty line decreases from 62.4 to 28.9 percent over the three years. Moreover, the percentage of households in the three villages living below the poverty line is smaller than the percentage of all households in Russia living below the poverty line (36.9 percent) in 1997 (*Rossiia* 1998:193). Analysis of changes in the income status of households in the three villages from 1995 to 1997 shows that 1.5 percent of households dropped into poverty, 75.6 percent did not change their income status, and 22.9 percent moved out of poverty.

Household Capital and Ownership of Durable Goods

The panel data provide support for hypothesis nine that higher levels of human and social capital will be associated with higher levels of material goods. An important indicator of the differentiation between households in the emerging market economy is the extent to which households with different levels of human and social capital are able to purchase durable goods, such as automobiles, VCRs, and telephones. A telephone, in itself, is not a very expensive good to purchase, even in rural Russia. However, installing a telephone line in a peasant household is extremely expensive, requiring a good deal of cash to be paid to appropriate authorities.

Table 9.11 shows that the most expensive durable goods, automobiles, are very strongly associated with the amount of labor in a household. As we have shown in Chapter 7 and earlier in this chapter, households with more labor are able to produce more and sell more from their household enterprises. In turn, this makes up a substantial portion of household income. In this regard, it is important to observe that over 80 percent of the households with the highest level of human capital have automobiles in 1997, compared with zero and 25 percent for households with the two lowest levels of human capital. The relationship between household labor and purchase of VCRs is also positive and statistically significant, although the relationship is not as strong as in the case of household labor and automobiles. The relationship between household labor and the ability of a household to purchase a telephone is somewhat stronger than in the case of household labor and VCRs.

Table 9.11: Mean Number of Durable Goods Owned by Household Labor in Three Russian Villages from 1995 to 1997

	Cars			Vidoes			Telephones		
	1995	1996	1997	1995	1996	1997	1995	1996	1997
0-1.74	0.0 (n=143)	0.0 (n=146)	0.0 (n=147)	0.0 (n=143)	0.0 (n=146)	0.0 (n=147)	0.0 (n=143)	0.0 (n=146)	0.0 (n=147)
1.75-2.74	0.18 (n=164)	0.25 (n=177)	0.25 (n=173)	0.10 (n=164)	0.14 (n=177)	0.22 (n=173)	0.18 (n=164)	0.21 (n=177)	0.19 (n=173)
2.75-3.74	0.28 (n=108)	0.38 (n=95)	0.42 (n=93)	0.14 (n=108)	0.23 (n=95)	0.36 (n=93)	0.17 (n=108)	0.26 (n=95)	0.30 (n=93)
3.75-4.74	0.35 (n=34)	0.46 (n=35)	0.36 (n=44)	0.12 (n=34)	0.20 (n=35)	0.34 (n=44)	0.18 (n=34)	0.23 (n=35)	0.30 (n=44)
4.75+	0.36 (n=14)	0.50 (n=10)	0.83 (n=6)	0.0 (n=14)	0.10 (n=10)	0.33 (n=6)	0.43 (n=14)	0.30 (n=10)	0.33 (n=6)
Total N=463	0.17	0.23	0.23	0.0	0.12	0.19	0.15	0.18	0.18

Cars by Household Labor Groups: 1995 $F(4)=11.24$, $p<.001$; Scheffe 1 x 2, $p<.01$, 1x 3, $p<.001$, 1 x 4, $p<.001$, 1 x 5, $p<.05$, 2 x 3, $p<.05$; 1996 $F(4)=15.21$, $p<.001$; Scheffe 1 x 2, $p<.001$, 1 x 3, $p<.001$, 1 x 4, $p<.001$, 1 x 5, $p<.01$, 2 x 3, $p<.05$; 1997 $F(4)=18.09$, $p<.001$; Scheffe 1 x 2, $p<.001$, 1 x 3, $p<.001$, 1 x 4, $p<.001$, 1 x 5, $p<.001$.

Videos by Household Labor Groups: 1995 $F(4)=2.46$, $p<.05$; Scheffe 1 x 3, $p<.08$; 1996 $F(4)=9.17$, $p<.001$; Scheffe 1 x 2, $p<.01$, 1 x 3, $p<.001$; 1997 $F(4)=16.78$, $p<.001$; Scheffe 1 x 2, $p<.001$, 1 x 3, $p<.001$, 1 x 4, $p<.001$.

Telephones by Household Labor Groups: 1995 $F(4)=5.33$, $p<.001$; Scheffe 1 x 2, $p<.05$, 1 x 5, $p<.01$; 1996 $F(4)=6.50$, $p<.001$; Scheffe 1 x 2, $p<.01$, 1 x 3, $p<.001$; 1997 $F(4)=6.88$, $p<.001$; Scheffe 1 x 3, $p<.001$, 1 x 5, $p<.05$.

Table 9.12 shows that the extent to which a household is involved in the institutional life of the village community is associated with purchases of durable goods. Households that have at least a moderate level of involvement in the ceremonies and festivals of other families and the village as a whole are more likely to have automobiles than are their neighbors who have no involvement in the social life of their community. Again, as noted with respect to household income, it appears that the most important distinction is between no involvement and moderate levels of involvement in the village community. Additional levels of involvement do not appear to bring as great a return to the household as the initial decision to at least participate occasionally in village and other household social activities.

Table 9.12: Mean Number of Durable Goods Owned by Level of Community Involvement in Three Russian Villages from 1995 to 1997

	Cars			Videos			Telephones		
	1995	1996	1997	1995	1996	1997	1995	1996	1997
1.00-2.99 (low)	0.0 (n=83)	0.0 (n=75)	0.0 (n=88)	0.0 (n=83)	0.0 (n=75)	0.0 (n=88)	0.0 (n=83)	0.0 (n=75)	0.10 (n=88)
3.00-5.99 (middle)	0.17 (n=326)	0.24 (n=346)	0.27 (n=355)	0.10 (n=83)	0.13 (n=346)	0.21 (n=355)	0.13 (n=326)	0.17 (n=346)	0.19 (n=355)
6.00-7.00 (high)	0.30 (n=54)	0.38 (n=42)	0.25 (n=20)	0.0 (n=83)	0.19 (n=42)	0.50 (n=20)	0.35 (n=54)	0.38 (n=42)	0.45 (n=20)
Total	0.17 (N=463)	0.23 (N=463)	0.23 (N=463)	0.0 (N=463)	0.12 (N=463)	0.19 (N=463)	0.15 (N=463)	0.17 (N=463)	0.18 (N=463)

Cars by Community Involvement, 1995 $F_{(2)}=4.83$, $p=.01$; Scheffe Low by High, $p<.01$. 1996 $F_{(2)}=6.14$, $p<.01$; Scheffe Low by Middle, $p<.05$, Low by High, $p<.01$. 1997 $F_{(2)}=5.34$, $p<.01$; Scheffe Low by Middle, $p<.01$.

Videos by Community Involvement, 1995 $F_{(2)}=1.79$, $p=$ns. 1996 $F_{(2)}=4.33$, $p<.05$; Scheffe Low by Middle, $p<.05$, Low x High, $p<.05$. 1997 $F_{(2)}=12.13$, $p<.001$; Scheffe Low by Middel, $p<.01$, Low x High, $p<.001$, Middle x High, $p<.01$.

Telephones by Community Involvement, 1995 $F_{(2)}=11.23$, $p=.001$; Scheffe Low by High, $p<.001$, Middle by High, $p<.01$; 1996 $F_{(2)}=8.34$, $p<.001$; Scheffe Low by High, $p<.001$, Middle by High, $p<.01$; 1997 $F_{(2)}=6.86$, $p<.001$, Scheffe Low by High, $p<.001$, Middle x High, $p<.01$.

There is no association between level of community involvement and owning VCRs in 1995, a modest relationship in 1996, but a strong relationship in 1997. There is a much stronger relationship between community involvement and ownership of telephones. In fact, this relationship is much stronger than the relationship between household labor and telephone ownership. This suggests that in this particular area, having contacts outside of the household is especially important in finding ways to work with local authorities to purchase telephone service.

The statistically significant, but modest, relationships between ownership of the three types of durable goods and the size of household helping networks is shown in Table 9.13. The relationship between network size and ownership of VCRs and telephones is curvilinear, following the pattern of relationships between this form of social capital and income reported earlier. The relationship between size of helping networks and ownership of automobiles does not follow the curvilinear pattern we have observed elsewhere. There is, however, a fairly simple explanation here. The relatively high level of vehicle ownership among those households with extremely high numbers of persons in their helping networks is an indicator of a household composed of elderly persons who have substantial support from relatives. Typically, these households have adult children living in cities who are able to provide the resources to assist these persons to purchase automobiles.

Table 9.13: Mean Number of Durable Goods Owned by Size of Non-Redundant Helping Networks in Three Russian Villages from 1995 to 1997

	Cars			Videos			Telephones		
	1995	1996	1997	1995	1996	1997	1995	1996	1997
0-3	0.0 (n=102)	0.11 (n=91)	0.05 (n=89)	0.0 (n=102)	0.02 (n=91)	0.06 (n=89)	0.0 (n=102)	0.03 (n=91)	0.01 (n=89)
4-5	0.19 (n=155)	0.18 (n=145)	0.18 (n=133)	0.0 (n=155)	0.08 (n=145)	0.11 (n=133)	0.12 (n=155)	0.16 (n=145)	0.01 (n=133)
6	0.20 (n=66)	0.28 (n=74)	0.30 (n=79)	0.14 (n=66)	0.14 (n=74)	0.24 (n=79)	0.29 (n=66)	0.23 (n=74)	0.34 (n=79)
7-10	0.21 (n=122)	0.31 (n=139)	0.34 (n=144)	0.11 (n=122)	0.22 (n=139)	0.32 (n=144)	0.16 (n=122)	0.25 (n=139)	0.26 (n=144)
11-17	0.33 (n=18)	0.36 (n=14)	0.33 (n=18)	0.28 (n=18)	0.07 (n=14)	0.17 (n=18)	0.33 (n=18)	0.21 (n=14)	0.17 (n=18)
Total	0.17 (N=463)	0.23 (N=463)	0.23 (N=463)	0.08 (N=463)	0.12 (N=463)	0.19 (N=463)	0.15 (N=463)	0.17 (N=463)	0.18 (N=463)

Cars by Nonredundant Helping Network Groups: 1995 $F(4)=3.27$, $p=.01$: Scheffe 1 x 4 $p=.10$, 1 x 5, $p<.11$; 1996 $F(4)=3.89$, $p=01$: Scheffe 1 x 4, $p=.05$; 1997 $F(4)=7.43$, $p=.001$; Scheffe 1 x 3, $p=.01$, 1 x 4, $p<.001$.
Videos by Nonredundant Helping Network Groups: 1995 $F(4)=3.54$, $p=.01$, Scheffe 1 x 5, $p<.05$; 1996 $F(4)=6.71$, $p=.001$, Scheffe 1 x 4, $p<.001$, 2 x 4, $p<.01$; 1997 $F(4)=8.66$, $p=.001$ Scheffe 1 x 3, $p<.05$, 1 x 4, $p<.001$, 2 x 4, $p<.001$.
Telephones by Nonredundant Helping Network Groups: 1995 $F(4)=6.79$, $p=.001$: Scheffe 1 x 3, $p=.001$, 1 x 5, $p<.05$; 1996 $F(4)=5.04$, $p=001$; Scheffe 1 x 3, $p<.05$, 1 x 4, $p=.01$; 1997 $F(4)=10.05$, $p=.000$: Scheffe 1 x 3, $p=.001$, 1 x 4, $p=.001$, 2 x 3, $p=.0o1$, 2 x 4, $p<.01$.

In summary, the most important finding is that a household's level of human labor is the best predictor of the kind of material conditions in which persons in that household will live. As we showed in Chapter 5 this form of human capital is embedded within the social organization of the household, which is a form of social capital. The "purer" forms of social capital, community involvement and social helping networks, also appear to assist families in purchasing these benefits of the market economy, however, their impact is much more modest than is the level of household labor. The critical issue here is whether it is possible to increase the economic returns from a household's investments in the purer forms of social capital. In order to provide material incentives for rural Russians to support economic and political reforms it would seem to be necessary to develop more ongoing cooperative institutions between households that produce mutual economic benefits for participants. This is a critical point to which we will return in Chapter 12.

Summary

The findings in this chapter show that the image of lack of change in the Russian countryside is simply wrong. We have shown that there has been considerable change in Russian villages and households in a relatively short period of time. Although the large enterprises continue to exist, they now must compete with other institutional sectors in the village for labor and resources. As a result, there has been a significant shift in employment patterns in the villages during the past three years. A significant minority of employed persons now work outside of the *kolkhozy* or *TOOs*. This is due to increased opportunities for employment in the public sector, local government, which is no longer totally dependent on the large enterprises.

Moreover, the majority of households now obtain the majority of their income, on average, close to two-thirds, from household enterprises. This is a fundamental structural change that has occurred in villages where only a few short years ago virtually all households were dependent on the large enterprises for the majority of their income.

Moreover, we have shown that despite the macro-economic figures that show a deterioration of material conditions in the Russian countryside, a substantial minority of households has purchased large durable goods items, such as automobiles and VCRs.

At the same time, however, our data also show that the degree of success in household adaptation to the market economy has varied considerably from one household to another. As predicted in hypotheses 8 and 9, households with higher levels of human and social capital, including household labor, community involvement and nonredundant social network ties, have been much more successful in obtaining income from a variety of sources and purchasing durable goods.

Finally, we have found that the economic "returns" to households from increasing their household labor is much greater than the returns they can receive from participating in the social life of their village community. Thus, although the social capital linkages between the household and others outside it, helping networks and community involvement, do have a positive relationship with material well being, that relationship is not nearly as powerful as what occurs within the household itself. This raises some important questions about the difficulties so far in inducing rural Russian households to become more involved in the institution building of a rural civic society.

10 Mental Health and Symptoms of Depression

Larry D. Dershem, Valeri V. Patsiorkovski and David J. O'Brien

Introduction

In the preceding chapters, both human and social capital have been shown to effect the material and economic status of rural households in Russia. Now we will examine the effect of human and social capital on a more intimate aspect of life closely associated with change, the psychological well being of individuals. Our expectation is that there will be a further differentiation in the psychological well being, specifically, depressive symptoms, of individuals by the level of human and social capital within their household (see Chapter 2). We expect that having more household labor during these times of economic crisis will instill a sense of security among household members, reduce the physical drudgery of all members and provide interpersonal emotional support. Any conditions that help with household production should also help individuals within the household deal with the psychological distress associated with persistent economic crisis and social change. Thus, individuals in households with more working age adult members should experience lower levels of stress in getting basic tasks completed that are related to agricultural production. Moreover, they will feel more secure about their ability to increase production in order to keep up with the increased demands of a nascent market economy. In these households, the benefits of having more working age adults will outweigh what are likely to be "transaction costs" associated with dealing with a larger number of interpersonal relationships. It is hypothesized that the amount of labor present in a respondent's household will have both direct and indirect negative effects on depression (hypothesis 10a).

Social exchange and helping networks have been found in other settings to facilitate coping with difficult situations, especially when these situations put increased demands on individuals for time, energy and material resources (Lin et al. 1986). Other studies have found that community attachment is positively related to mental health for rural residents in the American Midwest (O'Brien et al. 1994). Thus, we would expect that these forms of social capital would have a direct positive effect on the mental health of individuals in the household. In addition, the positive association between social capital and household production should produce an indirect positive effect on psychological mood in the household. As in the case of household labor, it is expected that there will be "transaction costs" associated with maintaining social networks and community attachments but that these costs will be offset by the positive benefits derived from these forms of social capital. Therefore, social networks and community attachment will have both direct and indirect negative effects on depression (hypothesis 11a). The panel study allows for the testing of these hypotheses, along with appropriate controls for demographic factors and depressive symptoms, such as negative and positive life events.

Economic Stress and Symptoms of Depression

Poor quality living conditions confronted by people in rural Russia during the Soviet era, and the pressure brought on by post-Soviet economic restructuring and weakened support of the social safety net by the federal government, lead to the expectation that a high percentage of rural Russians will show symptoms of depression. In a study of three Russian villages in 1993, 70 percent of the 207 respondents showed substantial symptoms of depression. Those who had the greatest number of depressive symptoms were the elderly, those in poor health, and women (Dershem et al 1996). The higher depressive symptoms found among those in poorer health and among women are similar to findings in other societies. However, the association between age and depression reflects a situation that is quite different from findings in Western nations. In studies in the U.S., for example, there is a negative association between age and depression until age 69 but depressive symptoms increase from age 70 onward (Gatz and Hurwicz 1990; O'Brien et al. 1994). In contrast, in the rural Russian study referred to earlier, symptoms of depression decreased until age 40 but began increasing again for those 41 year of age or older, and increased substantially for persons over 74 years of age (Dershem et al. 1996).

Anthropologists have suggested that fear, despair and fatalism permeate the peasant way of life. Poverty and marginalization, according to Baily (1966:405), lead peasants to have a negative view of life because,

> ...they see little security in their own life. No one can be sure whether the harvest will be good or bad: no one can be sure who will be alive this time next year, or next week. In two or three years a rich man can become poor or a poor man rich...In circumstances like this, no one can tell that man is the master of his environment.

According to Foster (1965), the peasant concept of "limited good" creates a pessimistic attitude that all good things exist in limited quantities and that there is no way to increase their availability. As a result, suspicion, vindictiveness, spite, envy, malicious gossip, and lack of cooperation mark interpersonal relationships with nonfamily members. Moreover, peasants believe leaders are instrumental and exploitive, the success of others means my failure, and the malevolence of others cause my failure.

Historically, rural Russians have experienced, during the past two centuries, extreme calamities, wars, revolutions, and epochs of political repression and physical depravation. Thus, everyday existence has often been insecure and unstable for the majority of the rural population. Most likely, these historical events have created a chronic and systematic level of stress among Russians living in the countryside. Economic stress, as Ross and Huber (1985) and Armstrong and Schulman (1990) found in the US, increases depression. As shown by Turner et al. (1995), experiencing historical and contemporary systemic stressors are relevant to understanding the ties between the social conditions of life and psychiatric disorders.

Description of CES-D Scale

Symptoms of depression were measured with a modified version of the Center for Epidemiological Studies Depression Scale (CES-D), a self-reported depressed-mood or symptom scale (Radloff 1977). Due to time and resource constraints, a modified version of the scale, which uses 11 of the 20 items from the original scale was used. The CES-D scale elicits symptoms of depression in general community settings during the week before the interview (Roberts 1980). Since this scale is not designed to identify diagnostic categories of mental illness it is inappropriate for a complete epidemiological survey. Rather, it attempts to measure groups of symptoms; depressed mood, worthlessness, hopelessness and fear.

A variety of survey research studies of general populations in rural areas have used the CES-D scale (Boyd et al. 1982; Husaini et al. 1980; Linn and Husaini 1987; Neff and Husaini 1987; Hoyt 1988; Ciarlo and Tweed 1992; Belyea and Lobao 1990; O'Brien et al. 1994). In addition, it has been used in cross-cultural studies of depression (Roberts 1982; Ying 1988). Studies of the CES-D scale indicate that it has very good internal consistency, acceptable test-retest stability, and construct validity in general urban and rural U.S. populations. Moreover, it is relatively easy to administer (Himmelfarb and Murrell 1983; Deforge and Sobal 1988; Kohout et al. 1993). The 1993 study in three rural Russian villages showed that the CES-D had 1) a relatively high response rate, 2) good internal consistency, 3) sensible factors and 4) reasonably high correlations with subjective quality of life assessments of various life domain measures. The 11 items comprising the scale clustered into three dimensions: depressed and somatic, positive and interpersonal (Dershem et al. 1996).

Table 10.1: English and Russian Versions* of the Modified Version of Center for Epidemiological Studies of Depression (CES-D) Scale

I'm going to read some statements which describe how people sometimes feel. Please tell me how often during the last week you felt this way: 3= Most times (5-7 days), 2= Moderate (3-4 days), 1= Sometimes (1-2 days), 0= Rarely (no days).

English Version	Russian Version
1. I felt that everything I did was an effort.	Все, что я делал, требовало больших усилий.
2. I felt fearful.	Мне было страшно.
3. I enjoyed life.	Я был доволен жизнью.
4. I felt lonely.	Я чувствовал себя одиноким.
5. I felt that I could not shake the blues.	Я не мог изъбавиться от чувства грусти.
6. I was happy.	Я был счастлив.
7. I could not get going.	Я не мог нечего делать, опустились руки.
8. I did not feel like eating; my appetite was poor.	Я не хотел есть, у меня был плохой аппетит.
9. I felt hopeful about the future.	Я был оптимистически настроен.
10. I felt depressed.	Я чувствовал депрессию.
11. I felt that people disliked me.	Я чувствовал, что люди нерасположены ко мне.

* Russian Version developed by the Institute for Socio-Economic Studies of Population, Russian Academy of Sciences. Because of cultural and language differences, the Russian version is not a literal translation of the English version.

The English and Russian versions of the modified CES-D scale are presented in Table 10.1. For analysis, the three positive items were reverse coded. To allow for comparison of the scores on the modified scale with the scores of persons who have responded to the full scale in previous studies, a conversion formula was employed. This formula [standard CES-D=(1.866 x modified CES-D) = 0.5318] transforms the scores from the truncated scale to what the numbers would have been if the full scale was used. Previous research has shown an alpha reliability of 0.75 for the modified scale and a correlation of 0.95 between responses to the truncated and full versions of the CES-D (O'Hara et al. 1985).

196 *Household Capital and the Agrarian Problem in Russia*

Table 10.2 shows the alpha reliability, mean, standard deviation, range and percentage of respondents above the depression criterion for the CES-D scale in 1995, 1996 and 1997. The alpha reliability coefficients of the CES-D scale are high for each year of the three-wave panel study (0.87, 0.85 and 0.88 respectively). These reliability coefficients are higher than the reliability coefficient for the CES-D scale in the 1993 study in rural Russia (0.79), and for general populations in the U.S.

Table 10.2: Descriptive Statistics of the Modified CES-D Scale in Rural Russia

Descriptive statistics of modified CES-D scale	Rural Russia			
	Preliminary study (n=207)		Panel Study (n=463)	
	1993	1995	1996	1997
Alpha reliability	0.79	0.87	0.85	0.88
Mean	22.45	22.70	22.51	21.92
Standard Deviation	10.52	10.62	10.19	10.43
Hi	48.12	56.51	58.38	56.51
Low	3.33	2.40	2.40	2.40
% above depression criterion of 16	70.0	67.0	68.9	65.0

Approximately two-thirds of all respondents had CES-D scores above the criterion for showing symptoms of depression (16 or greater) over the three years. This figure is quite high compared with 12 percent of the respondents in rural Tennessee in 1979 (Husaini et al.), 26.1 percent of the respondents in rural Missouri in 1993 (O'Brien et al. 1994) and 35 percent of the respondents in rural Ohio in 1987 (Belyea and Lobao 1990).

Factor analysis was used to divide the CES-D into specific dimensions of depression. These dimensions (shown in Appendix 12) are (1) depressed affect, (2) somatic-retarded activity, (3) interpersonal relations, and (4) positive affect. With one exception, this factor structure is similar to the factor structure of the CES-D found in other studies (Ensel 1986:60). In previous factor analyses, the depressed item loads with the depressed factor, whereas in this study, it crosses over to the somatic-retarded activity factor. A detailed interpretation of these factors is beyond the focus of this chapter, however, it is reasonable to conclude that the internal structure of the CES-D scale is consistently and reliably factored into a set of four depressive categories. (See Dershem et al. 1996 for more details.)

Since the CES-D scale is designed to detect depressed mood during the week before the interview, the likelihood of change occurring from one year to the next is highly probable. Ensel (1986:63) reported a Pearson correlation of 0.41 of the CES-D scale at two points in time over a period of one year among respondents. The overtime correlations of the CES-D scale in the panel study were 0.74 from 1995 to 1996, 0.77 from 1996 to 1997, and 0.63 from 1995 to 1997. These relatively high correlations overtime indicate the endurance of depressive mood states in these respondents.

Symptoms of Depression by Year and Village

The mean CES-D score by village for each year of the panel study is shown in Table 10.3. Latonovo, located in the southern black earth area of southern Russia, had the lowest average CES-D scores of all three villages, while Sviattsovo, located in north central Russia, had the highest averages from 1995 to 1997. The average CES-D score in Latonovo was significantly lower than the average CES-D score in Vengerovka and Sviattsovo in 1995. However, by 1996 and 1997 the average CES-D score in Vengerovka had dropped by almost 10 percent, leaving only the average CES-D score in Sviattsovo significantly higher than the average CES-D score in Latonovo. Thus, the primary difference between CES-D scores is between the two villages in the southern and central portion of European Russian and the village in the north. Several factors may explain this village level difference. First, there is a longer growing season in the south than in the north, which aids household agricultural production and sales in the former. Increased food production and sales translate into a less stress related to feeding household members and income. Second, elderly people, who are more numerous in the northern part of the Central Region than in the southern regions of Russia, tend to have more symptoms of depression than younger people. Third, the settlements in the North tend to be sparsely populated, leaving people more isolated than in settlements in the south.

Table 10.3: Mean CES-D Score by Village and Year

Village	1995	1996	1997
Latonovo (n=157)	20.27	21.01	20.08
Vengerovka (n=156)	24.20	22.37	22.35
Sviattsovo (n=150)	23.67	24.22	23.84
Total (N=463)	22.70	22.51	21.92

CES-D by Village by Year: 1995 F(2)=6.43, p<.01; Latonovo by Vengerovka, p<.01, Latonovo by Sviattsovo, p<.05. 1996 F92)=3.87, p<.05; Latonovo by Sviattsovo, p<.05. 1997 F(2)=4.10, p<.05; Latonovo by Sviattsovo, p<.05.

Household Labor, Social Capital and Symptoms of Depression

As hypothesized (H-10), symptoms of depression decrease as the number of working age adults in the household increase who can assist with household production.

Table 10.4 shows that respondents living in households with the lowest amount of household labor had higher average CES-D scores than all other household labor groups. In 1996 and 1997 the effect is linear. For each increase in the amount of household labor there is a decrease in the average CES-D score.

It is also interesting to note that in 1995 the group with the highest amount of household labor had an average CES-D score 40 percent lower than the group with the lowest amount of household labor. By 1997, this difference increased. That is, the average CES-D score for the group with the highest amount of household labor was 54 percent lower than the average CES-D score for the group with the lowest amount of household labor. Thus, additional household members, who are advantageous to increased production for subsistence and survival in economic crisis, reduce the number of depressive symptoms experienced by individuals in those households.

Mental Health and Symptoms of Depression 199

Table 10.4: Mean CES-D Score by Household Labor from 1995 to 1997

Weighted Number of Adults in Household	1995	1996	1997
0-1.74	29.49 (n=143)	28.22 (n146)	28.32 (n=147)
1.75-2.74	20.50 (n=164)	21.13 (n=177)	19.97 (n=173)
2.75-3.74	18.21 (n=108)	18.45 (n=95)	17.85 (n=93)
3.75-4.74	21.06 (n=34)	18.07 (n=35)	18.05 (n=44)
4.75+	17.73 (n=14)	17.70 (n=10)	12.66 (n=6)
Total (N=463)	22.70	22.51	21.92

1995 $F(4)=27.18$, $p<.001$; 1 by 2, $p<.001$, 1 x 3, $p<.001$, 1 by 4, $p<.001$, 1 x 5, $p<.001$. 1996 $F(4)=21.47$, $p<.001$; 1 by 2, $p<.001$, 1 x 3, $p<.001$, 1 by 4, $p<.001$, 1 x 5, $p<.05$. 1997 $F(4)=26.36$, $p<.001$; 1 by 2, $p<.001$, 1 x 3, $p<.001$, 1 by 4, $p<.001$, 1 x 5, $p<.01$.

Mental health, in these villages, is moderately related to community attachment, as measured by the sense of "fit" in the village a respondent feels (see Table 10.5). Respondents who reported feeling a high degree of attachment to their village had a lower average CES-D score than respondents who felt low to middle levels of attachment to their village. Community involvement was not related to mental health (not shown).

Table 10.5: Mean CES-D Score by Year and Sense of Fit in the Village

Sense of "Fit" in village*	Mean CES-D Score		
	1995	1996	1997
Low and Middle	23.18 (n=367)	22.81 (n=348)	22.09 (n=361)
High	20.86 (n=96)	21.61 (n=115)	21.31 (n=102)
Total (N=463)	22.70	22.51	21.92

* Low and Middle were combined due to Low having only 5 observations.
1995: $F(1)=3.63$, $p<.06$; 1996 $F(1)=1.91$, $p=ns$; 1997 $F(1)=0.44$, $P=ns$.

As mentioned earlier, social exchange and helping networks have been found in other settings to facilitate coping with difficult situations, especially when these situations put increased demands on individuals for time, energy and material resources, especially outside the household. The average size of these nonredundant personal helping networks increased slightly over the three-year period, from 5.5 people in 1995, to 5.6 in 1996 and 5.8 people in 1997. These helping networks were comprised of

200 *Household Capital and the Agrarian Problem in Russia*

individuals who provided the respondent with labor (assistance with the private plot) and socio-emotional support (discuss important matters).

Similarly, as shown in Table 10.6 and as hypothesized (H-11), as the size of personal helping networks increased the average CES-D score decreased in 1995. However, in 1996 and 1997 the relationship became nonlinear. Those respondents with small and large personal networks had higher CES-D scores than respondents with middle size helping networks. As shown earlier (Chapters 6 and 8), large helping networks were associated with households consisting of elderly dependent persons. Thus, the size of a household's helping network had a curvilinear relationship with depression, as well as with production and income from salary and wages.

Table 10.6: Mean CES-D Score by Number of Non-Redundant Ties

Size of Helping Network	1995	1996	1997
0-3	27.92 (n=102)	28.03 (n=91)	29.13 (n=89)
4-5	22.78 (n=155)	21.88 (n=145)	22.35 (n=133)
6	22.73 (n=66)	21.89 (n=74)	19.64 (n=79)
7-10	18.90 (n=122)	19.74 (n=139)	18.40 (n=144)
11-17	18.05 (n=18)	23.86 (n=14)	21.16 (n=18)
Total (N=463)	22.70	22.51	21.92

1995 $F_{(4)}=11.95$, $p<.001$; 1 x 2, $p<.01$, 1 x 3, $p<.05$, 1 x 4, $p<.001$, 1 x 5, $p<.01$. 1996 $F_{(4)}=10.26$, $p<.001$; 1 x 2, $p<.01$, 1 x 3, $p<.01$, 1 x 4, $p<.001$. 1997 $F_{(4)}=18.10$; 1 x 2, $p<.001$, 1 x 3, $p<.001$, 1 x 4, $p<.001$, 1 x 5, $p<.05$, 2 x 4, $p<.05$.

Additional Influences on Depression

The focus of this chapter is on the effect of social capital on symptoms of depression. However, research has shown that many other conditions and factors influence a person's psychological state, such as age, marital status, health, negative and positive life events and household income. Some of these additional indicators are adaptations of existing measures (e.g., health status) and others were constructed specifically for this study (e.g., negative and positive life events) (Ensel 1986).

Health Satisfaction, Marital Status and Age

A self-evaluation scale was used to measure health status (Maddox and Douglas 1973) in which respondents were asked, "On a scale to 1 to 7, with 1 being dissatisfied and 7 being satisfied, how satisfied are you with your health?" Overall, respondents had a consistently low level of satisfaction with their health or an average of 3.5 in all three years of the study. Self-rated health reflects perceived health, which is important for both well-being and health service utilization and has been shown to predict mortality (Appels et al. 1996; Mossey and Shapiro 1982).

The marital statuses were single (i.e., never married), married, divorced/separated and widow(er). Each marital status was dichotomized into a dummy variable with married as the reference group.

Age has been found in other studies to have a nonlinear relationship with depression, with CES-D scores decreasing up to age 69 and then increasing from 70 onward. Thus a quadratic term was included (Gatz and Hurwicz 1990). Age is measured in number of years. In 1995, the average age of respondents was 51.7 years. Since this is a panel study of the same individuals, the average age of respondents increased by one year for 1996 and 1997.

Negative Life Events

Life events and experiences have been shown to increase stress (Pearlin 1989). Stressful life events may have direct and indirect affects on depression. To tap stressful life events respondents were asked if in the last year they, or a household member, (1) lost a job, (2) changed job for worse one, (3) did not receive salary or pension, (4) divorced/separated, (5) had something stolen, (6) were physically attacked, (7) had a relative or a friend with a serious illness, (8) a family member with a serious illness or injury, (9) had someone in the household who died, (10) had someone who moved out of the respondent's home and the respondent did not want them to. The sum total of these ten life events formed a Negative Life Event index. The average number of negative life events reported in each year remained virtually constant, slightly less than two.

Positive Life Events

There are other types of life events that can lead to decreased levels of stress. Respondents were asked if in the last year they, or a household member, (1) received a good job position, (2) a family member or friend had married, (3) built or bought new house, (4) had a child or grandchild born, (5) had someone move out of the home who the respondent wanted to move out, (6) purchased something they have wanted for a long time. The sum total of these six life events formed a Positive Life Event index. On average, respondents reported slightly less than one positive life event in each of the three years.

Modeling the Effect of Household Capital on Depression

Figure 10.1 illustrates the structural model of the effect of household labor, personal helping networks and sense of "fit" into the village on symptoms of depression. The findings are organized into two sections: (1) general effects of control variables and (2) the direct, indirect and mediating effects of social capital (personal helping networks and sense of "fit" into the village) on symptoms of depression.

General Effects of Control Variables

In general, all control variables had associations with symptoms of depression similar to the association of comparable variables found in other studies. Using the magnitude of the standardized beta coefficients to rank the effect of the control variables, the strongest effects on symptoms of depression were the levels of a respondent's health satisfaction, his or her level of monthly household income, marital status, and gender.

Respondents who reported higher levels of satisfaction with their health had lower CES-D scores. On a seven-point scale with 1 as dissatisfied and 7 as satisfied, respondents reporting a high level of satisfaction with their health (6 or 7) had an average CES-D score of 15.72, which is almost two times lower than the average CES-D score of 30.78 for respondents who reported low levels of health satisfaction (1 or 2). Respondents who were neither dissatisfied nor satisfied with their health (3 to 5) had an average CES-D score of 20.63.

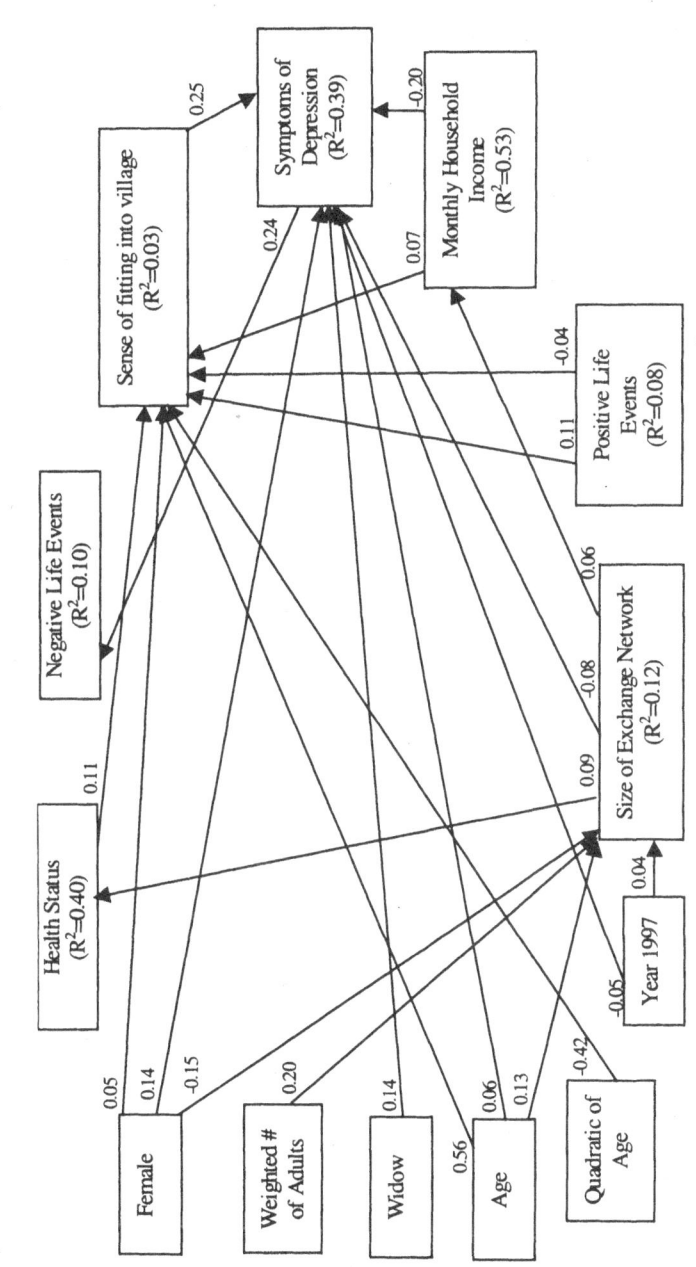

Figure 10.1: Modeling Household Capital and Symptoms of Depression in the Three Russian Villages ($X^2=35.54$, $df=28$, $p=.16$, $GFI=.996$, $AGFI=987$, all paths are of statistical significance of 0.05 or greater)

Table 10.7: Standardized Regression Coefficients for Observed Exogenous Variables and Explained Variance (R^2) for Structural Model of Depression

Observed endogenous	R^2	Standardized regression weights	Observed exogenous
CES-D	0.39		
		-0.259	Health status
		-0.198	Monthly household income
		0.139	Widow
		-0.139	Sense of "fit" in village
		0.138	Female
		-0.082	Size of helping network
		0.064	Respondent's age
		-0.049	Year 1997 compared to 1995
		-0.044	Positive life events
Size of helping network	0.12		
		0.204	Weighted number of adults
		-0.153	Female
		-0.126	Respondent's age
		0.044	Year 1997 compared to 1995
Sense of "fit" in village	0.03		
		0.558	Respondent's age
		-0.424	Quadratic of age
		0.113	Positive life events
		0.107	Health status
		0.070	Monthly household income
		0.051	Female
Monthly household income	0.53		
		0.557	Weighted number of adults
		-0.134	Year 1997 compared to 1995
		-0.222	Widow
		0.064	Size of helping network
		-0.058	Female
Health status	0.40		
		-1.126	Respondent's age
		0.594	Quadratic of age
		-0.102	Negative life events
		0.086	Size of helping network
		0.085	Monthly household income
		-0.055	Female
Negative life events	0.11		
		-0.083	Respondent's age
		0.230	Positive life events
		0.238	CES-D
		0.143	Sense of "fit" in village
Positive life events	0.08		
		0.218	Monthly household income
		-0.134	Respondent's age
		0.100	Size of helping network
		-0.096	Weighted number of adults
		0.048	Year 1997 compared to 1995
		0.063	Female

$X^2=35.54$, $df=28$, $p=.16$, $GFI=.996$, $AGFI=987$.

Mental Health and Symptoms of Depression 205

The economic condition of the household affected symptoms of depression. Graph 10.1 presents the average CES-D score by ten equal income groups.[1] The average CES-D score (32.54) is very high for respondents living in households earning less than 204,000 rubles per month. The sharpest decline in average CES-D scores occurs between respondents living in households with income from 203,783 to 286,044 rubles per month (group 2) and respondents living in household with income from 450,001 to 525,140 rubles per month (group 5). Once household income reaches approximately 597,644 rubles per month, average CES-D scores decrease slightly although steadily.

Graph 10.1: Mean CES-D Score by Income

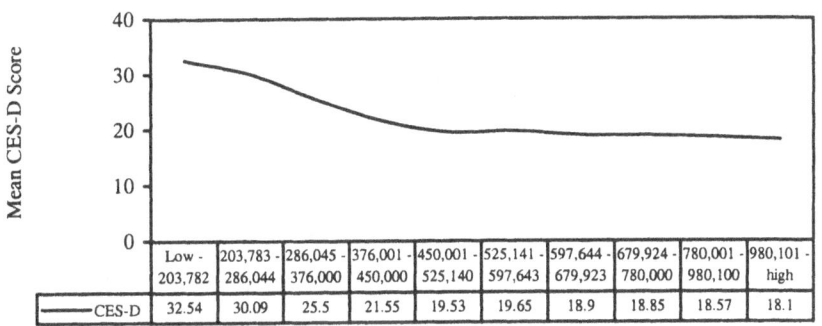

Monthly Household Income

Higher levels of depression were found among widows compared to respondents who were married. The average CES-D score for widows was 29.73 and 18.79 for married respondents. The average CES-D score for respondents who are divorced is quite high (25.34) but due to their small numbers (less than 4 percent of respondents) the difference in scores was not statistically significant.

---

[1] See Chapter 8 on why household income was used and not per capita income.

206 Household Capital and the Agrarian Problem in Russia

Women reported significantly more symptoms of depression than men, which is commonly found in studies of depression in other countries (Weissman and Lerman 1977; O'Hara et al. 1985). The average CES-D score for women (23.80) was 23 percent higher than the average CES-D score for men (18.22). Overall, the average CES-D score decreased 3.4 percent between 1995 (22.70) and 1997 (21.92).

Effects of Social Capital on Symptoms of Depression

Figure 10.1 presents the structural model of the effects of social capital - personal exchange networks and sense of "fitting" into the village - on symptoms of depression. Other significant paths are not shown but are reported in Table 10.7.

Direct Effects of Social Capital

The structural equation model illustrated that the size of helping networks was directly and negatively associated with symptoms of depression. That is, as the size of a person's helping networks increased his or her mean CES-D depression score decreased.

Graph 10.2: Mean CES-D Scale Score and Percentage of Respondents Above Depression Criterion by Size of Helping Network

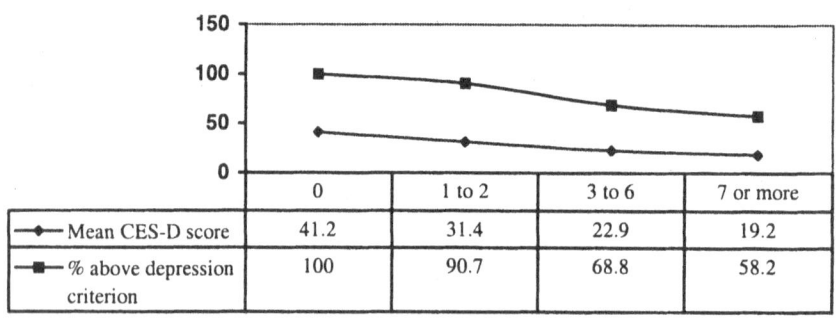

	0	1 to 2	3 to 6	7 or more
Mean CES-D score	41.2	31.4	22.9	19.2
% above depression criterion	100	90.7	68.8	58.2

Size of Helping Network

Graph 10.2 illustrates this association. Respondents who reported no one with whom they could rely on for help (n=5) had a mean CES-D score of 41.2, decreasing to 31.4 for respondents with one to two people, 22.9 for respondents with three to six people and 19.2 for respondents with seven or more people.

Graph 10.2 also shows the percentage of respondents above the CES-D scale criterion for symptoms of depression by size of helping network. All of the respondents who reported no personal helping network had CES-D scores above the depression criterion, decreasing to 90.7 percent of respondents who had 1 or 2 people, 68.8 percent for respondents with 3 to 6 people, and 58.2 percent of respondents reporting seven or more people in their personal helping network.

The second measure of social capital, a respondent's sense of "fitting" into the village, is also directly and negatively associated with symptoms of depression. Respondents who felt a close attachment to their village reported fewer symptoms of depression. These findings provide support for hypothesis 11a that social networks and community attachment would have a negative effects on depression.

Indirect Effects of Social Capital

Personal helping networks also were indirectly associated with lower symptoms of depression through positive life events, health status and household income (see Table 10.8). The total causal effect of personal helping networks on symptoms of depression is 0.13, almost equal to the direct effect of a respondent's sense of "fitting" into the village (-0.14).

Table 10.8: Direct, Indirect and Total Causal Effects of Personal Helping Networks on Symptoms of Depression*

Causal Effects	CES-D
Direct	-0.08
Indirect	
Positive Life Events	-0.01
Health Status	-0.02
Household Income	0.02
Total	0.13

*The indirect causal effect of personal networks on symptoms of depression through positive life events equals (0.10 x -0.05) = -0.01.

Mediating Effects of Social Capital

In addition to their direct and indirect effects, personal helping networks also play a mediating role between household labor, gender, age and year, and symptoms of depression. As depicted in Figure 10.1, personal helping networks, in addition to their direct and indirect effects, play a mediating role between household labor, gender, age and year on symptoms of depression.

The strongest mediating role of personal helping networks on symptoms of depression is through household labor. Graph 10.3 shows the mean CES-D score by household labor groups and size of helping network. Personal helping networks have the greatest effect on decreasing CES-D scores for respondents in the lowest household labor group (0 to 1.74), which are primarily single-person households. Single person households reporting no helping network had an average CES-D score of 41.2. Single person households with seven or more members in their helping network, however, had an average CES-D score of 23.4. For respondents living in two person household labor groups (1.75-2.74 and 3.75-4.74), three or more members in their personal networks seems to be the threshold for significantly decreasing CES-D scores. These two household labor groups are primarily retired couples and two generation extended family households, respectively.

Mental Health and Symptoms of Depression 209

Graph 10.3: The Relationship Between Household Labor Groups and Symptoms of Depression as Mediated by the Size of the Personal Helping Networks

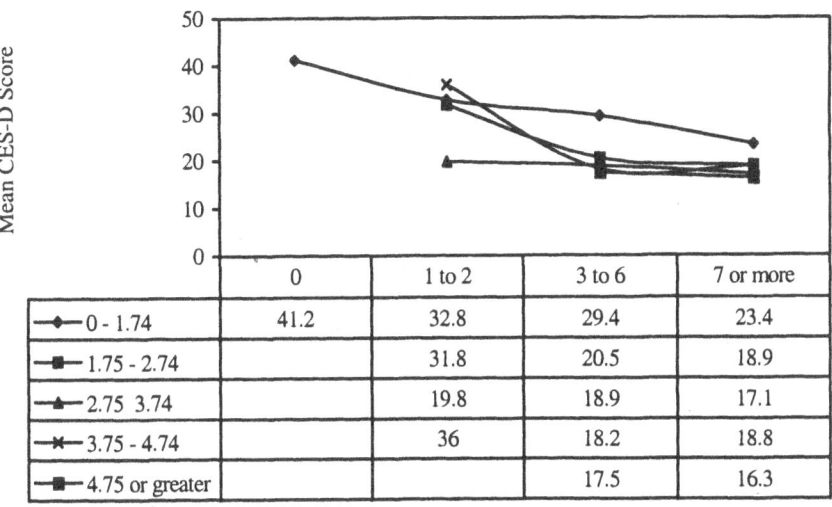

Size of Helping Network

The size of personal helping networks had little effect on decreasing the CES-D scores for respondents living in the middle household labor group (2.75-3.74), which consists primarily of employed couples with children, and respondents living in the highest household labor group (4.75 or higher), which consists primarily of large (three generation) extended families. This shows that during rapid socio-economic change a respondent's level of stress and depression is affected by certain combinations of household labor and social capital. Respondents living in households with lower levels of household labor, such as single persons, retired couples and two generation extended families, who have three or more members in their personal network experience less stress and depression than respondents living in similar households with small or nonexistent personal helping networks. However, respondents living in households comprised of only their spouse and one or two children, or large three generation extended family households, experience lower levels of stress and depression regardless of the size of their personal helping network. These findings provide tentative support for hypothesis 10a. Household labor has an indirect effect through helping networks rather than a direct effect on depression.

The size of personal helping networks reduces symptoms of depression for both men and women, as shown in Graph 10.4, but decreases symptoms of depression primarily for women. The mean CES-D score for women reporting no personal helping network was 41.2, decreasing to 33.3 for women with one or two members, 24.2 for women reporting three to six members and 20.2 for women reporting seven or more members.

Graph 10.4: The Relationship Between Gender and Symptoms of Depression as Mediated by the Size of Personal Helping Networks

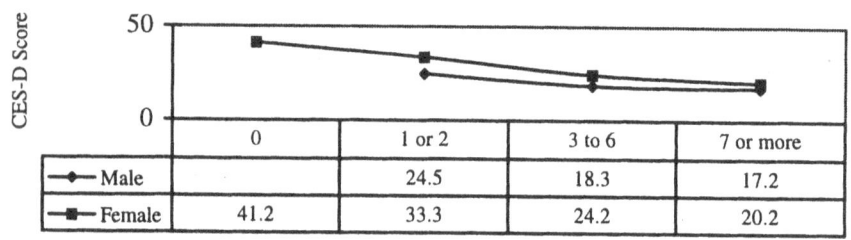

Size of Helping Network

The last mediating relationship of personal helping networks on symptoms of depression is age. Respondents were grouped into six age categories. Graph 10.5 shows the mean CES-D score by the size of personal helping networks for each age group. The mean CES-D score decreases within each age group as the size of personal helping network increases. The sharpest decline in CES-D scores by size of helping network is for respondents between 50 and 59 years of age. For respondents 70 years of age or older, the decline in CES-D scores occurs for those with more than three or more members in their helping network. The effect of helping networks on CES-D scores is almost identical for respondents who are 30 to 49 years of age.

Mental Health and Symptoms of Depression 211

Graph 10.5: The Relationship Between Age and Symptoms of Depression as Mediated by the Size of Personal Helping Networks

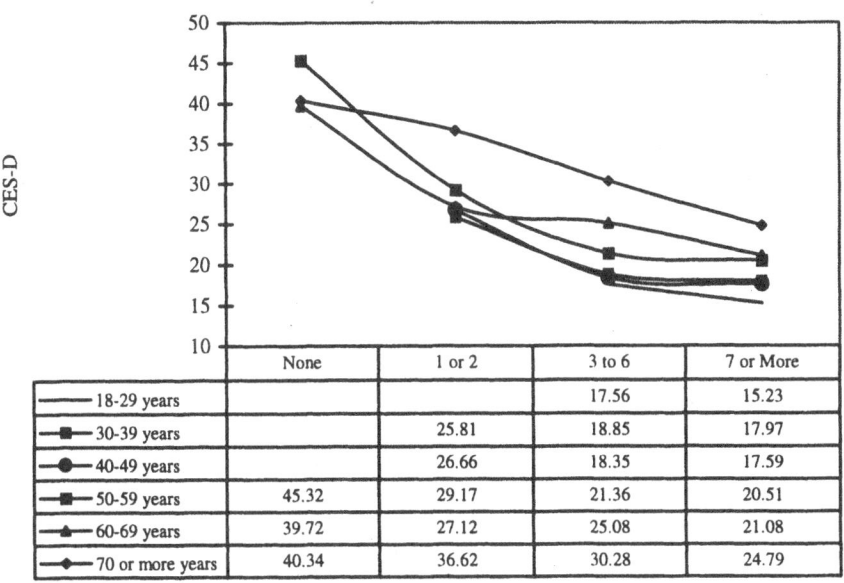

	None	1 or 2	3 to 6	7 or More
—— 18-29 years			17.56	15.23
—■— 30-39 years		25.81	18.85	17.97
—●— 40-49 years		26.66	18.35	17.59
—■— 50-59 years	45.32	29.17	21.36	20.51
—▲— 60-69 years	39.72	27.12	25.08	21.08
—◆— 70 or more years	40.34	36.62	30.28	24.79

Size of Helping Network

Summary

Rural Russians are going through a period of rapid socio-economic change, which can lead to increased psychological stress and depression. However, as shown in this chapter, two forms of social capital; helping networks and sense of "fitting" into the community had a direct effect on decreasing symptoms of depression.

In addition, helping networks had indirect effects on symptoms of depression by increasing positive life events, a respondent's evaluation of his or her health, and the household's monthly income.

Finally, helping networks also moderated the effect of various social and economic characteristics of respondents on symptoms of depression. Larger size helping networks decreased symptoms of depression regardless of the amount of labor in the household. Larger size helping networks also decreased symptoms of depression for women and for all age groups, especially those who were 50 to 59 years of age.

11 Subjective Quality of Life

David J. O'Brien, Valeri V. Patsiorkovski and Larry D. Dershem

Introduction

Much has been written, in both the Russian and Western popular press, about the quality of life in Russia after the collapse of the Soviet Union. There has been much speculation, for example, about the reaction of Russian citizens to the collapse of the command economy and the introduction of a market economy. A special concern of these musings has been the political reaction of ordinary Russians to the high inflation of the early post-Soviet period, the loss of the traditional social service safety net, and the new demands for households to compete in a market economy. More often than not, however, the "experts" have not been very accurate about the reactions of Russians to the difficulties presented by the collapse of the traditional economic system and the emergence of a changing and still ill defined new socio-economic system. There has not been mass starvation, nor have a majority of voters sought a return to Soviet Communism. The widely discussed fears about a fascist nationalism and/or a brown-red coalition have not materialized. Rather, as noted in Chapter 1 of this book, ordinary Russians have been quite creative in adapting to the uncertainties of the new situation. At the same time, of course, public opinion polls have shown strong negative assessments of the effectiveness of the Russian government (White et al. 1997).

The primary reason for the survival of Russian rural households during this transitional period has been their ability to increase production for consumption and additional income. This has resulted in the development of different types of household enterprises that operate outside official government regulations and taxes. These survival strategies are based upon the development of different types of household capital. Especially important here is household labor, which is human capital *embedded* in household relations, a type of social capital (Chapters 3 and 5), and community involvement and social exchange helping networks, which are purer forms of social capital (Chapter 6). As shown in Chapters 8 and 9, these enterprises represent an incremental, but nonetheless substantial structural change in the economic base of the Russian countryside.

This is admittedly a more hopeful view of Russian agrarian life than is typically presented in Western scholarly writings. There are, however, three crucial caveats that have been identified in the analysis presented in previous chapters. First, as shown in Chapters 8 and 9, there is an uneven adaptation to the new economy. Households with more household capital are increasing production and new sources of income more rapidly than households with lower amounts of household capital. Second, there are substantial psychological costs that households are experiencing as they struggle to make a living within the very difficult environment of labor-intensive agriculture. Third, levels of psychological distress vary considerably by levels of household capital, especially the extent to which households possess effective social exchange helping networks (see Chapter 10).

The focus of this chapter will be on how different levels of household capital affect the way in which rural Russians *perceive* the conditions of their lives in this transitional period. An individual's evaluation of his or her life obviously will be affected by material conditions and the extent to which he or she is experiencing psychological distress in trying to cope with those conditions. But a subjective evaluation, also involves an individual's *comparative* measurement of his or her life (or some part of that life) against some kind of standard. Thus, for example, a young professional person may be experiencing a great deal of stress while he or she is building a career and yet feel that these difficulties are worthwhile in the long-run because eventually they will produce a successful career, with all of its attendant benefits. Contrariwise, an older professional may not be experiencing much "pressure" at all in his or her work, and receiving an excellent salary, but feel that his or her work is not fulfilling because he or she is stuck in some sort of "dead-end" position. In short, subjective quality of life depends to significant degree on the way in which individuals *feel* about these material conditions. Campbell, Converse and Rodgers note that:

> ...the relationship between objective conditions and psychological states is very imperfect and...in order to know the quality of life experience it will be necessary to go directly to the individual himself for his description of how his life *feels* to him (our emphasis) (1976:4).

The way in which individuals perceive the quality of their lives is critical in scientific assessments of how they are affected by social change. Public opinion research, however, shows that individuals do not evaluate their lives in a simple uni-dimensional way. Rather, different dimensions of an individual's life are affected differently by social change. A change that may improve an evaluation of one area of life may, at the same time, decrease personal evaluation of other areas of life. Most important, individuals differ in the personal characteristics and resources they possess with which to deal with social change and this will have an important impact on how they evaluate different areas of their lives.

The subjective quality of life model developed by Campbell, Converse and Rodgers (1976) permits us to analyze specific *paths* through which various types of household capital affect an individual's overall quality of life in the Russian countryside today. The model assumes that overall quality of life is made up of satisfaction with areas or "domains" of life. Thus, different levels of household capital may affect overall life satisfaction through specific life domains.

The general conceptual approach of Campbell, Converse and Rodgers will be employed to understand how rural Russians perceive the quality of their lives for each year of the panel study. Findings for this analysis, as in the previous chapter, are based on the responses of the individuals who are informants for the household. This is a limitation, but it nonetheless provides a basic insight into how ordinary Russians perceive their lives. Later, a structural equation model will be used to measure the relationships between specific types of household capital, life domains and life in general.

Satisfaction with Life Domains and Life in General

In each year of the panel study, respondents were asked to assess the level of satisfaction with seven life domains and their life in general. Two of these life domains, health and employment satisfactions are not presented in this analysis. Health is not included because it is highly correlated with age. Employment is not included because a large proportion (approximately one-third) of respondents are not employed, which would greatly reduce the sample size. The five domains that were analyzed represent concerns that had different degrees of immediacy in a person's life. Satisfaction with one's current marital status, family life, and income were assumed to be the most immediate concerns. Next in order of immediacy would be satisfaction with the village and then with the overall situation in the country.

Measurement of different levels of satisfaction with each of the five domains and life in general were obtained from responses to three pairs of semantic differential seven-point scales. Respondents were asked to identify the extent to which they found each of the domains and life in general to be "dissatisfied...satisfied," "unpleasant...pleasant," and "disappointing ...rewarding." There were a total of 44 missing cases on the different quality of life items in the panel data, which resulted in a loss of 132 out of 1389 cases in the pooled data set for a total of 1257 cases. The alpha reliabilities of the indexes created by combining the three semantic differential items for the six life domains and life-as-a-whole were 0.91 or higher during all three years of the panel study (for more details on measurement see Chapter 4).

The mean levels of satisfaction of the total sample for the five life domains and life in general from 1995 to 1997 are shown in Table 11.1. Despite popular press reports, rural Russians evaluated their overall quality of life about average, that is, neither completely satisfied nor dissatisfied. Moreover, there was a slight increase in the mean level of overall quality of life satisfaction over the three-year period, although this increase is not statistically significant. Over the three-year panel study, only family and marital status satisfactions had mean levels of satisfaction above the average. Only village satisfaction decreased over the three years, even though its mean level remained above average.

For each of the three years of the study, respondents were twice as satisfied with their family, marital status and village, and 50 percent more satisfied with their income, as they were with the situation in the country. Despite fluctuations, the large difference between the mean levels of satisfaction with interpersonal relations and economic and political structures remains constant over the three years. This indicates that, overall, Russian villagers have higher levels of satisfaction with micro-level relations than with macro-level conditions in the country.

Subjective Quality of Life 217

Table 11.1: Mean Level of Satisfaction with Five Life Domains and Life in General* by Year

Domain satisfaction	1995 (n=419)	1996 (n=419)	1997 (n=419)
Marital status	4.33	4.45	4.42
Family	4.71	4.79	4.80
Income	2.85	3.07	3.07
Village	4.39	4.04	3.92
Country	1.97	1.93	2.05
Life in general	3.51	3.54	3.58

*Scale: 1=dissatisfied to 7=satisfied.
Marital status by Year $F(2)=0.82$, p<ns: Family by Year $F(2)=0.62$, Income by Year; $F(2)=6.12$, p<.01; Scheffe 1995 by 1996 p<.01, 1995 by 1997 p<.01: Village by Year; $F(2)=24.86$, p<.000; Scheffe 1995 by 1996 p<.000, 1995 by 1997 p<.000: Country by Year $F(2)=3.38$, p<.05, 1996 by 1997 p<.05; Life in General by Year: $F(2)=0.69$, p<ns.

A comparison of the mean levels of satisfaction with different life domains for each of the three years of the panel study shows that important changes occurred in a relatively brief period of time. Satisfaction with the most intimate types of interpersonal relations in the institutions of marriage and family remained constant during the three years, reflecting their normative commitment and capacity to support individuals during the current crisis. There were, however, significant changes in satisfaction with income, village and the overall situation in the country. An increased optimism about material conditions was represented by an increase in satisfaction with income and the situation in the country, but this was offset by a substantial decrease in level of satisfaction with the village.

The increases in satisfaction with income and the situation in the country is explained in large measure by the successful adaptation strategies developed by households, especially the growth of household enterprises. These strategies have produced some visible material gains for a substantial number of households in the three villages (see Chapters 8 and 9).

In our view, the most important reason for the lower assessment of village life is that the villages in the study, like most Russian villages, have not adapted their social, economic, and political institutions to the realities of a market economy as fast as have the households. While households have managed to develop and expand interpersonal helping networks to increase the effectiveness of their enterprises, the villages have not developed new institutional structures to assist households in purchasing inputs, processing raw agricultural commodities, or in marketing these products. In general, the villages have not developed many important institutions of civic society, such as profit and not-for-profit organizations (see Chapter 12), and have not found ways to replace the social service infrastructure support once obtained from the *kolkhozy* and *sovkhozy*.

Subjective Quality of Life 219

Table 11.2: Mean Level of Satisfaction with Five Life Domains and Life in General by Year and Village

Domain Satisfaction	1995			1996			1997		
	Latonovo (n=139)	Vengerovka (n=146)	Sviattsovo (n=134)	Latonovo (n=139)	Vengerovka (n=146)	Sviattsovo (n=134)	Latonovo (n=139)	Vengerovka (n=146)	Sviattsovo (n=134)
Marital status	4.76	4.20	4.05	4.77	4.22	4.38	4.74	4.14	4.40
Family	4.96	4.56	4.62	4.99	4.57	4.82	4.93	4.68	4.79
Income	2.71	3.07	2.76	3.04	3.02	3.17	3.04	3.18	2.99
Village	4.29	4.49	4.40	3.69	4.29	4.14	3.77	4.11	3.86
Country	1.92	2.03	1.96	1.96	1.99	1.83	2.05	2.06	2.05
Life in general	3.49	3.57	3.46	3.51	3.66	3.45	3.56	3.67	3.50

*Scale: 1=dissatisfied to 7=satisfied.

1995: Marital status by village $F(2)=9.06$, $p<.001$: Scheffe Latonovo by Vengerovka $p<.01$, Latonovo by Sviattsovo $p<.000$; Family by village $F(2)=4.13$, $p<.05$: Scheffe Latonovo by Vengerovka $p<.05$; Income by village $F(2)=4.52$, $p<.05$; Scheffe Latonovo by Vengerovka $p<.05$: Village by village $F(2)=1.29$, $p=$ns: Country by village $F(2)=0.82$ $p=$ns.: Life in general by village $F(2)=0.52$, $p=$ns.

1996: Marital status by village $F(2)=5.56$, $p<.01$; Scheffe Latonovo by Vengerovka $p<.01$: Family by village $F(2)=4.37$ $p<.01$: Scheffe Latonovo by Vengerovka $p<.01$; Income by village $F(2)=0.70$, $p=$ns: Village by village $F(2)=13.10$ $p<.000$: Scheffe Latonovo by Vengerovka $p<.000$, Latonovo by Sviattsovo $p<.001$: Country by village $F(2)=2.27$ $p=$ns: Life in general by village $F(2)=2.57$, $p=$ns.

1997: Marital status by village $F(2)=7.68$, $p<.001$; Scheffe Latonovo by Vengerovka $p<.001$: Family by village $F(2)=1.67$ $p=$ns: Income by village $F(2)=1.59$, $p=$ns: Village by village $F(2)=4.86$, $p<.01$; Scheffe Latonovo by Vengerovka $p<.01$: Country by village $F(2)=0.00$ $p=$ns: Life in general by village $F(2)=1.68$, $p=$ns.

220 *Household Capital and the Agrarian Problem in Russia*

Table 11.2 shows the mean level of satisfaction with the five life domains and life in general in each of the three villages in the study. There are no differences between the villages in mean levels of overall life satisfaction, but there are some important differences between them with respect to domain satisfaction. Latonovo has a higher level of satisfaction with marital status than does Vengerovka in each of the three years, reflecting a higher percentage of married persons in the former than in the latter; 72.4 percent compared to 63.0 percent in 1995. Satisfaction with family is also higher in Latonovo than in Vengerovka in 1995 and in 1996, but the differences between these villages disappears in 1997. Respondents in Vengerovka have a higher level of satisfaction with income than do respondents in Latonovo in 1995, reflecting the higher levels of household income in the latter village during that year (see Chapter 8). Differences in satisfaction with income disappear, however, in 1996 and 1997.

Residents of Vengerovka have significantly higher levels of satisfaction with their village than do residents of Latonovo in 1996 and in 1997. These differences reflect the fact that the village administration and the *TOO* has been much more supportive of household enterprises in Vengerovka than in Latonovo. As noted earlier, the chairman of the *TOO* in Latonovo eliminated a crucial type of material support for sales of household dairy products in Latonovo in 1996, the year in which the level of village satisfaction is the lowest in that village (see Chapter 8). In addition, 1996 is the year in which there is the greatest difference in mean levels of village satisfaction between Latonovo and Vengerovka.

Household Capital and Satisfaction with Life Domains and Life in General

The central thesis of this study has been that household capital plays the crucial role in determining the extent to which families are able to adapt to the exigencies of an emerging market economy. In this section, we will analyze four aspects of household capital, household labor, community involvement, sense of fit, and helping networks. Physical capital (land, animals and tools) is not included in this analysis because of the overwhelming complexity of the model.

We have seen so far that households with more household labor have higher levels of agricultural production (Chapter 8) and higher overall

Subjective Quality of Life 221

incomes (Chapter 9), as well as lower levels of symptoms of depression (Chapter 10).

Table 11.3 shows the relationship between levels of household labor and satisfaction with the five specific life domains and life in general. The strongest associations with household labor are marital and family satisfaction. It will be shown later in Figure 11.1 and Figure 11.2 that the basic differentiation here is between those who are married and those who are not. Persons with higher levels of household labor are more likely to be married and, in turn, to have a more positive view of their marital status and their family than are those who are divorced, widowed or who have never been married.

The association between level of household labor and family satisfaction is complicated by the strong relationship between family and marital satisfaction. This will be discussed in more detail below when we examine the full model showing relationships between all variables that affect subjective quality of life. At this juncture, however, it is important to note that the relationship between level of household labor and satisfaction with marital status and family highlights the importance of household labor in the economic and psychological adaptation of families to market relations. Thus, not only does household labor have a strong and direct effect on the ability of the household to produce agricultural commodities, or develop some other type of household enterprise, but this type of capital also increases positive feelings about life in the household itself.

Table 11.3: Mean Level of Satisfaction with Four Life Domains and Life in General by Year and Household Labor

Weighted Number of Adults	Marital status	Family	Income	Village	Country	Life in general
1995						
0-1.74 (n=117)	3.14	4.15	2.75	4.20	2.13	3.40
1.75-2.74 (n=153)	4.56	4.76	2.94	4.40	2.00	3.54
2.75-3.74 (n=103)	5.04	5.10	2.86	4.51	1.85	3.55
3.75-4.74 (n=32)	4.96	5.02	2.65	4.72	1.82	3.64
4.75 + (n=14)	5.14	5.40	3.21	4.36	1.57	3.50
Total (N=419)	4.33	4.71	2.85	4.39	1.97	3.51
1996						
0-1.74 (n=119)	3.45	4.35	2.94	4.08	2.03	3.43
1.75-2.74 (n=166)	4.77	4.89	3.02	4.00	1.91	3.60
2.75-3.74 (n=92)	4.96	5.07	3.22	4.10	1.87	3.59
3.75-4.74 (n=32)	5.11	5.17	3.16	3.99	1.85	3.51
4.75 + (n=10)	4.40	4.50	3.87	3.90	1.83	3.63
Total (N=419)	4.45	4.79	3.07	4.04	1.93	3.54
1997						
0-1.74 (n=119)	3.47	4.46	2.84	3.95	2.13	3.44
1.75-2.74 (n=163)	4.74	4.84	3.10	3.82	2.07	3.58
2.75-3.74 (n=90)	4.76	4.93	3.20	3.99	1.98	3.70
3.75-4.74 (n=41)	5.00	5.20	3.22	3.94	1.95	3.68
4.75 + (n=6)	5.67	5.61	4.00	4.39	1.89	3.83
Total (N=419)	4.42	4.80	3.07	3.92	2.05	3.58

Table 11.3 continued

*Scale: 1=dissatisfied to 7=satisfied.

1995: Marital status by Household Labor $F(4)=37.68$, $p<.000$; Scheffe 1 x 2 $p<.000$, 1 x 3 $p<.000$, 1 x 4 $p<.000$, 1 x 5 $p<.000$, 2 x 3 $p<.000$, 2 x 4 $p<.01$: Family by Household Labor $F(4)=10.36$ $p<.000$: Scheffe 1 x 2 $p<.000$, 1 x 3 $p<.01$, 1 x 4 $p<.000$, 1 x 5 $p<.01$; Income by Household Labor $F(4)=1.16$, $p=$ns: Village by Household Labor $F(4)=2.09$, $p=$ns.: Country by Household labor $F(4)=3.35$ $p.01$: Scheffe 1 x 3 $p<.10$: Life in General by Household Labor $F(4)=0.58$, $p=$ns.

1996: Marital status by Household Labor $F(4)=27.56$, $p<.000$; Scheffe 1 x 2 $p<.000$, 1 x 3 $p<.000$, 1 x 4 $p<.000$: Family by Household labor $F(4)=6.90$: Scheffe 1 x 2 $p<.000$, 1 x 3 $p<.000$, 1 x 4 $p<.000$; Income by Household Labor $F(4)=2.23$, $p=$ns: Village by Household Labor $F(4)=0.23$, $p=$ns: Country by Household labor $F(4)=1.01$ $p=$ns; Life in General by Household Labor $F(4)=0.93$, $p=$ns.

1997: Marital status by Household Labor $F(4)=28.81$, $p<.000$; Scheffe 1 x 2 $p<.000$, 1 x 3 $p<.000$, 1 x 4 $p<.000$, 1 x 5 $p<.001$: Family by Household labor $F(4)=4.85$ $p<.001$: Scheffe 1 x 4 $p<.05$; Income by Household Labor $F(4)=3.99$, $p<.01$; Scheffe 1 x 3 $p<.08$: Village by Household Labor $F(4)=0.93$, $p=$ns: Country by Household labor $F(4)=0.99$ $p=$ns; Life in General by Household Labor $F(4)=1.87$, $p=$ns

Household labor is positively associated with satisfaction with income in 1997 and country in 1995. We will see later, however, that these relationships disappear when controlling for the effect of other variables on satisfaction with these domains. Alternatively, there are no significant zero-order relationships between household labor, on the one hand and village satisfaction and satisfaction with life in general, on the other, but relationships between these variables are statistically significant in the full structural equation model shown below (see below, Figure 11.1, Table 11.7). Overall, these findings provide support for hypothesis 10b that higher levels of household labor are associated with higher subjective evaluations of quality of life. These findings also support hypothesis 11b that "purer" forms of social capital have even stronger positive effects on Russian villages' evaluations of the quality of their lives. This is consistent with findings reported in the last chapter that social capital had a strong effect on mental health.

Households that are more involved in the village community (through participation in festivals and celebrations) have higher sales of agricultural products (Chapter 8) and higher overall incomes (Chapter 9). Respondents with a greater sense of "fit" in the village had fewer symptoms of depression (Chapter 10). The critical questions are: to what extent do these forms of household capital directly impact overall subjective life satisfaction and, to what extent do they work indirectly on life satisfaction through association with specific life domains? As shown in previous chapters, each type of household capital is somewhat different and serves a different function.

Community involvement is a behavioral measure of community attachment, measuring the extent to which individuals actually participate in village festivals and ceremonies and the ceremonies and festivals of other families. Thus, in addition to the satisfaction that is derived from participating in the institutional life of the community, community involvement has a clear instrumental function that enhances a household's ability to cope with a market economy.

Sense of community fit, on the other hand, measures a rural resident's sense of personal attachment to the village and its citizens. A sense of fit in the village increases the perception of personal security and thus individuals are better able to deal with the emotional strain connected with economic and social change.

Subjective Quality of Life 225

Table 11.4: Mean Level of Satisfaction with Four Life Domains and Life in General* by Year and Level of Community Involvement

Community Involvement	Marital status	Family	Income	Village	Country	Life in general
1995						
Low (n=78)	3.83	4.39	2.64	4.10	1.96	3.27
Middle (n=291)	4.35	4.70	2.87	4.35	2.01	3.47
High (n=50)	5.05	5.30	3.07	5.09	1.75	4.07
Total (N=419)	4.33	4.71	2.85	4.39	1.97	3.51
1996						
Low (n=65)	3.89	4.47	2.65	3.86	1.73	3.09
Middle (n=316)	4.51	4.81	3.16	3.99	1.97	3.58
High (n=38)	4.93	5.12	3.03	4.81	1.90	3.97
Total (N=419)	4.45	4.79	3.07	4.04	1.93	3.54
1997						
Low (n=78)	3.74	4.24	2.70	3.62	2.02	3.17
Middle (n=325)	4.56	4.92	3.16	3.95	2.05	3.65
High (n=16)	4.92	5.08	3.06	4.73	2.27	4.15
Total (N=419)	4.42	4.80	3.07	3.92	2.05	3.58

*Scale: 1=dissatisfied to 7=satisfied.

1995: Marital status by Community Involvement $F(2)=10.83$, $p<.000$; Scheffe Low by Middle $p<.000$, Low by High $p<.000$, Middle by High $p<.05$, Low by High $p<.01$: Family by Community Involvement $F(2)=7.96$, $p<.000$; Scheffe Low by High $p<.000$, Middle by High $p<.01$; Income by Community Involvement $F(2)=2.58$, $p=ns$; Village by Community Involvement $F(2)=15.20$, $p<.000$: Scheffe Low by High $p<.000$, Middle by High $p<.000$: Country by Community Involvement $F(2)=2.74$, $p=ns$; Life in General by Community Involvement $F(2)=11.34$, $p<.000$; Scheffe Low by Middle $p<.000$, Middle by High $p<.000$.

1996: Marital status by Community Involvement $F(2)=7.74$, $p<.000$; Scheffe Low by Middle $p<.01$, Low by High $p<.000$: Family by Community Involvement $F(2)=3.93$, $p<.05$: Scheffe Low by High $p<.05$; Income by Community Involvement $F(2)=5.80$, $p<.01$; Scheffe Low by Middle $p<.01$; Village by Community Involvement $F(2)=12.01$, $p<.000$: Scheffe Low by High $p<.000$, Middle by High $p<.000$: Country by Community Involvement $F(2)=3.64$, $p<.05$: Scheffe Low by Middle $p<.05$; Life in General by Community Involvement $F(2)=16.61$, $p=.000$; 1 x 2 $p<.07$; Scheffe Low by Middle $p<.000$, Low by High $p<.000$, Middle by High $p<.05$.

1997: Marital status by Community Involvement $F(2)=14.15$, $p<.000$: Scheffe Low by Middle $p<.000$, Low by High $p<.01$: Family by Community Involvement $F(2)=11.42$, $p<.000$: Scheffe Low by Middle $p<.000$, Low by High $p<.05$; Income by Community Involvement $F(2)=6.67$, $p<.01$; Scheffe Low by Middle $p<.01$: Village by Community Involvement $F(2)=10.28$, $p<.000$; Scheffe Low by Middle $p<.01$, Low by High $p<.000$, Middle by High $p<.01$: Country by Community Involvement $F(2)=0.92$, $p=ns$; Life in General by Community Involvement $F(2)=17.26$, $p<.000$; Scheffe Low by Middle $p<.000$, Low by High $p<.000$.

226 *Household Capital and the Agrarian Problem in Russia*

The associations between satisfaction with life domains and life in general and community involvement and sense of community fit are shown in Table 11.4 and Table 11.5. Not surprisingly, both community involvement and sense of fit are positively associated with village satisfaction. In addition, both types of community attachment are positively associated with satisfaction with life in general, although, as we will see later (Figure 11.1, Table 11.7), only community involvement is associated with overall life satisfaction in the structural equation model.

Most important, the two ways of relating to the village community have effects on different life domains, and, in one instance they have opposite effects on the same domain. Table 11.4 and Table 11.5 show that both types of community attachment are positively associated with marital and family satisfaction. In the structural equation model presented below, however, only sense of community fit has a relationship with marital and family satisfaction. This is consistent with the notion that sense of community fit is strongly associated with family life in rural Russia and thus is an important factor in reducing stress in the household. Likewise, persons with the highest sense of fitting in their village communities have more positive evaluations of their income and the situation in the country.

Subjective Quality of Life 227

Table 11.5: Mean Level of Satisfaction with Four Life Domains and Life in General* by Year and Sense of Fit in the Village

Sense of "Fit" In village**	Marital status	Family	Income	Village	Country	Life in general
			1995			
Low and Middle (n=333)	4.21	4.59	2.77	4.27	1.95	3.39
High (n=86)	4.81	5.19	3.15	4.86	2.07	3.97
Total (N=419)	4.33	4.71	2.85	4.39	1.97	3.51
			1996			
Low and Middle (n=315)	4.34	4.68	3.01	3.98	1.85	3.47
High (n=104)	4.79	5.13	3.27	4.23	2.16	3.77
Total (N=419)	4.45	4.79	3.07	4.04	1.93	3.54
			1997			
Low and Middle (n=)	4.35	4.73	3.02	3.81	1.99	3.49
High (n=)	4.66	5.04	3.26	4.29	2.28	3.90
Total (N=419)	4.42	4.80	3.07	3.92	2.05	3.58

*Scale: 1=do not fit to 7=fit.
** Low and Middle were combined due to the Low group having only 5 cases.
1995: Fit by Marital status $F(1)=11.39$, $p<.001$; Fit by Family $F(1)=15.45$, $p<.000$; Fit by Income $F(1)=8.14$, $p<.01$, Fit by Village $F(1)=22.55$, $p<.000$; Fit by Country $F(1)=1.81$, p=ns; Fit by Life in General $F(1)=25.95$, $p<.000$.
1996: Fit by Marital status $F(1)=7.71$, $p<.01$; Fit by Family $F(1)=11.09$, $p<.001$; Fit by Income $F(1)=4.28$, $p<.05$, Fit by Village $F(1)=4.30$, $p<.05$; Fit by Country $F(1)=16.76$, $p<.000$; Fit by Life in General $F(1)=11.39$, $p<.001$.
1997: Fit by Marital status $F(1)=4.00$, $p<.05$; Fit by Family $F(1)=5.11$, $p<.05$; Fit by Income $F(1)=4.40$, $p<.05$, Fit by Village $F(1)=19.22$, $p<.000$; Fit by Country $F(1)=13.15$, $p<.000$; Fit by Life in General $F(1)=20.11$, $p<.000$

228 Household Capital and the Agrarian Problem in Russia

Contrariwise, persons with the highest level of community involvement actually have less satisfaction with their income. It will be recalled that community involvement is primarily associated with agricultural sales and thus persons in households with more community involvement are likely to have the more successful enterprises in their villages. Their dissatisfaction with their incomes reflects the desire to increase their sales even further.

As with household labor, larger social helping networks (up to a point) have higher levels of agricultural production (Chapter 8) and higher overall incomes (Chapter 9), as well as lower levels of symptoms of depression (Chapter 10).

Table 11.6 shows overall positive relationships between the size of helping networks and satisfaction with marital status, family, and income. The relationships between size of helping networks and satisfaction with marital status and family are not linear. This reflects the pattern, described earlier in Chapters 6 and 8, where extremely high numbers of persons in a respondent's overall helping networks occurs mainly among elderly persons. Thus, the types of networks that are most effective in facilitating household production are those which are large, but not too large.

Size of helping networks is associated with satisfaction with village in 1995 and life in general in 1997. Nevertheless, the overall relationships between these variables during the three year course of the panel study do not hold up in the structural equation model that is described below (see Figure 11.1, Table 11.7).

Subjective Quality of Life 229

Table 11.6: Mean Level of Satisfaction with Four Life Domains and Life in General by Year and Size of Helping Network

Size of Helping Network	Marital status	Family	Income	Village	Country	Life in general
1995						
0-3 (n=79)	3.51	4.00	2.53	4.03	2.17	3.33
4-5 (n=146)	4.35	4.73	2.74	4.39	1.98	3.51
6 (n=64)	4.30	4.70	3.18	4.46	2.07	3.57
7-10 (n=116)	4.81	5.10	2.96	4.58	1.78	3.60
11-17 (n=14)	4.98	5.45	3.33	4.67	1.88	3.48
Total (N=419)	4.33	4.71	2.85	4.39	1.97	3.51
1996						
0-3 (n=70)	3.85	4.26	2.75	3.91	2.22	3.45
4-5 (n=129)	4.35	4.75	2.88	4.00	1.84	3.55
5 (n=74)	4.52	4.83	3.19	4.00	1.99	3.52
7-10 (n=133)	4.89	5.13	3.35	4.20	1.86	3.62
11-17 (n=13)	3.87	4.31	3.28	3.85	1.64	3.26
Total (N=419)	4.45	4.79	3.07	4.04	1.93	3.54
1997						
0-3 (n=69)	3.83	4.31	2.73	3.82	2.28	3.37
4-5 (n=117)	4.25	4.75	2.99	3.94	2.09	3.54
6 (n=78)	4.68	4.89	3.17	3.97	2.17	3.77
7-10 (n=139)	4.68	5.01	3.26	3.94	1.93	3.64
11-17 (n=16)	4.75	4.94	3.02	3.71	1.31	3.23
Total (N=419)	4.42	4.80	3.07	3.92	2.05	3.58

*Scale: 1=dissatisfied to 7=satisfied.
1995: Marital status by Helping Network F(4)=10.69, p<.000; Scheffe 1 x 2 p<.001, 1 xy 3 p<.05, 1 x 4 p<.000, 1 x 5 p<.01: Family by Helping Network F(4)=10.78, p<.000: Scheffe 1 by 2 p<.001, 1 x 3 p<.05, 1 by 4 p<.000, 1 by 5 p<.01; Income by Helping Network F(4)=4.59, p<.001: Scheffe 1 x 3 p<.01: Village by Helping Network F(4)=3.70, p<.000; Scheffe 1 x 4 p<.01: Country by Helping Network F(4)=3.72, p<.01: Scheffe 1 x 4 p<.01: Life in General by Helping Network F(4)=0.97, p=ns.
1996: Marital status by Helping Network F(4)=8.76, p<.000; Scheffe 1 x 4 p<.01, 2 x 4 p<.05: Family by Helping Network F(4)=7.56, p<.000: Scheffe 1 x 3 p<.000: Income by Helping Network F(4)=4.89, p<.001; Scheffe 1 x 4 p<.01, 2 x 4 p<.05: Village by Helping Network F(4)=1.78, p=ns: Country by Helping Network F(4)=5.11, p<.001: Scheffe 1 x 2 p<.01, 1 x 4 p<.01: Life in General by Helping Network F(4)=0.97, p=ns.
1997: Marital status by Helping Network F(4)=6.67, p<.000; Scheffe 1 x 3 p<.01, 1 x 4 p<.001: Family by Helping Network F(4)=4.44, p<.01: Income by Helping Network F(4)=4.03, p<.01; Scheffe 1 x 4 p<.01: Village by Helping Network F(4)=0.47, p=ns: Country by Helping Network F(4)=9.21, p<.000: Scheffe 1 x 4 p<.01, 1 x 5 p<.000, 2 x 5 p<.001, 3 x 5 p<.000, 4 x 5 p<.01: Life in General by Helping Network F(4)=3.58, p<.01: Scheffe 1 x 3 p<.05

It is important to note that there is a consistent negative relationship between size of helping networks and satisfaction with country. This relationship, coupled with the earlier observation that size of helping networks is positively associated with satisfaction with income, can be interpreted quite simply. Households that have larger networks are more effective in producing agricultural commodities (Chapter 8) and this translates into higher overall household income (Chapter 9). But, these same households are also the most likely to be frustrated with the lack of local and national institutional change that hinders expansion of their household enterprises. Typically, these more successful enterprises have reached the limit of what they can produce with the land and credit available to them. Central and regional governmental inaction in land reform and opportunities for household credit are basic causes of these limitations.

Modeling the Effects of Household Capital on Subjective Quality of Life

Figure 11.1 and Table 11.7 show the relationships between the various types of household capital, domain satisfactions and satisfaction with life in general. These relationships provide strong support for hypothesis 10b and 11b. Two out of the four measures of household capital, household labor and community involvement, have *direct* effects on overall life satisfaction. Both of these types of household capital play a critical role in household survival strategies and household enterprises.

Household labor has a strong positive association with marital satisfaction. As noted earlier, this is a bi-modal effect, with those who are married having more household labor than those who are not married (the latter being, largely, elderly widows). Although there is a positive zero-order relationship between household labor and family satisfaction, this relationship is reversed when we control for the effect of satisfaction with marital status. Those respondents who are married are more satisfied with a larger number of immediate family members living in the household. In contrast, for respondents who are not married (i.e., widow/er or divorced) the larger number of extended kin in the household entails more economic than emotional relations.

Household labor also has a direct negative relationship with village satisfaction. As noted earlier, this reflects the fact that households with more labor are frustrated because of the lack of institutional change at the village level; i.e., lack of land and lack of support for expansion of processing and marketing.

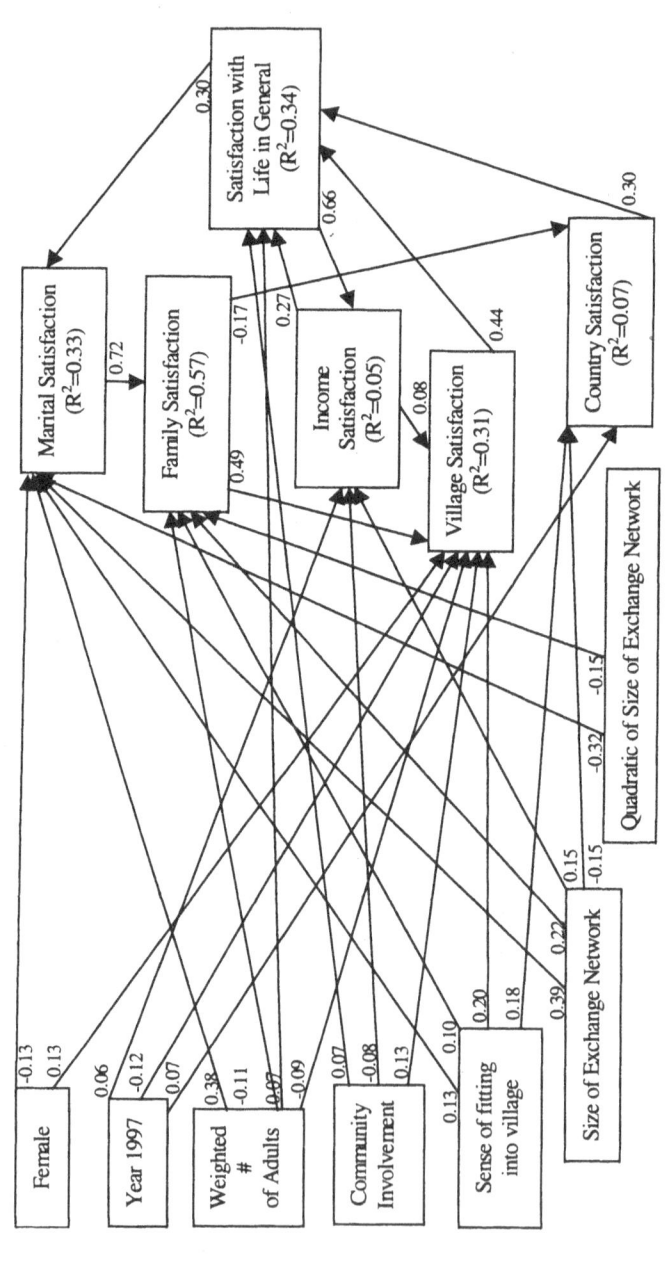

Figure 11.1: Effects of Household Capital on Five Quality of Life Domains and Life in General ($X^2=41.00$, df=30, p<.09, GFI=0.99, AGFI=0.99)

Subjective Quality of Life 233

Table 11.7: Standardized Regression Coefficients for Observed Exogenous Variables and Explained Variance (R^2) for Structural Model of Peasant Household Quality of Life

Observed endogenous	R^2	Standardized Regression weights	Observed exogenous
Life in General	0.34		
		0.44	Village satisfaction
		0.30	Country satisfaction
		0.27	Income satisfaction
		0.07	Community Involvement
		0.07	Weighted number of adults
LIFE DOMAINS			
Family Satisfaction	0.57		
		0.72	Marital status satisfaction
		0.22	Size of helping network
		-0.14	Quadratic of size of helping network
		-0.11	Weighted number of adults
		0.10	Sense of "fit" into the village
Marital Satisfaction	0.33		
		0.39	Size of helping network
		-0.32	Quadratic of size of helping network
		0.38	Weighted number of adults
		0.20	Overall satisfaction with Life
		-0.13	Female
		0.13	Sense of "fit" into the village
Village Satisfaction	0.31		
		0.49	Family satisfaction
		0.20	Sense of "fit" into the village
		0.13	Community Involvement
		0.13	Female
		-0.12	Year 1997
		-0.09	Weighted number of adults
		-0.08	Income satisfaction
Country Satisfaction	0.07		
		0.18	Sense of "fit" into the village
		-0.17	Family satisfaction
		-0.15	Size of helping network
		0.07	Year 1997
Income Satisfaction	0.05		
		0.66	Overall Life Satisfaction
		0.15	Size of helping network
		-0.08	Community Involvement
		0.06	Year 1997

$X^2=41.00$, $df=30$, $p<.09$, $GFI=0.99$, $AGFI=0.99$.

Community involvement has a direct positive effect on overall life satisfaction. It has an indirect positive effect on overall life satisfaction through its association with village satisfaction. In addition, community involvement has a negative indirect effect on overall life satisfaction through its association with income satisfaction. This reflects the fact that households with higher levels of community involvement had higher levels of sales and thus are frustrated at the constraints on expanding their household enterprises (see Chapter 8).

A respondent's sense of fitting into the community has a direct positive association with satisfaction with the situation in the country. A sense of fit is an indicator of a more *affective* type of social capital in which the attachment of individuals to the village community reduces their sense of anxiety about the current economic situation in which they are placed (see Chapter 10). Fitting into one's community affects subjective evaluations of life in general through its support for the various local institutions with which the rural Russian resident is involved; marriage, family and village.

Socio-emotional support derived from fitting into the village community makes it easier for rural households to cope with the difficult adjustments they have to go through during this transitional period. Alternatively, community involvement, which is more *instrumental* in nature, is related to expanding opportunities for sales of household goods and this would seem to lead to pressure for greater institutional change in order to increase household enterprise output.

Social support helping networks do not have a direct effect on overall life satisfaction, but they have indirect effects operating through four of the five life domains. Helping networks have a positive relationship with three of these four domains: marital status, family and village. This relationship is curvilinear with marital status and family satisfaction. Moderate numbers of persons in helping networks is associated with higher levels of satisfaction with marital status and family, while higher levels of help have a negative effect. Again, this reflects the fact that extremely large social exchange helping networks are associated with elderly, often single person, households.

The positive paths going from year 1997 to income and country satisfaction reflect the improvement in household enterprise output (Chapter 8) and household income (Chapter 9) in 1997. The negative path from year 1997 to village satisfaction reflects the growing frustration of rural households with the failure of village institutions to develop to meet the growing needs of household enterprises to function in a market economy.

Life in general has a causal effect on two of the five life domains, income and marital status. The effect of overall life satisfaction on income satisfaction is strong, and the relationship between these two variables is the only reciprocal relationship in the structural model. The effect of overall life satisfaction on income satisfaction is approximately 2.5 times stronger than the effect of income on overall life satisfaction. This suggests that global evaluations of life satisfaction, comprised of evaluations of both material conditions and social relations, have a stronger impact on evaluation of economic means than vice versa.

Summary

The findings in this chapter show clearly that levels of household capital not only affect the material conditions of Russian rural peasants but also have substantial effects on their evaluation of the quality of their lives. While overall subjective quality of life evaluations did not change significantly during the course of the panel study, there were important substantive changes in levels of satisfaction with various life domains. The material conditions of certain rural households improved during the course of the panel study and this is reflected in higher overall evaluations of income and country. At the same time, however, frustration of peasant households with a lack of institutional change at the local level, produced much lower levels of satisfaction with their villages. The higher levels of satisfaction by Vengerovka residents with their village reflects the fact that this village has been more progressive than the other two in making institutional changes, especially in providing credit for peasant household enterprise expansion. For the total sample in all three villages, however, the positive and negative changes in different domains have resulted in a lack of any shift in overall life satisfaction.

Differences between the specific types of human and social capital that we have been considering throughout this book become clearer once we examine their direct and indirect effects on overall life satisfaction. Household labor and personal helping networks are most closely associated with household production (Chapter 8). Moreover, they have the most important indirect effects on life satisfaction. Together, the amount of household labor and the size of helping networks indicate the capacity of the household as a production unit.

Level of community involvement is not an indicator of household production potential, but it does indicate the extent to which a household can earn income through sales of what it does produce. Thus, its effect on overall life satisfaction operates primarily through its association with income satisfaction. The negative relationship between community involvement and income satisfaction reflects the situation where households with the highest levels of sales are also the most frustrated with the failure of institutional change to support their desire to increase sales further.

Sense of fitting into the village community is a type of social capital that functions primarily as a source of emotional support during troubled times, operating through all five domains. This reflects the existential attachment to the village which so many writers, both Russian and non-Russian, have observed.

12 Supporting Sustainable Households and Communities in an Era of Globalization

David J. O'Brien, Valeri V. Patsiorkovski and Larry D. Dershem

Introduction

The problems facing rural households and villages in Russia ultimately must be framed within the constraints imposed by the globalization of the world economy and agriculture. The overriding trend in world agriculture is the reduction and eventual elimination of subsidies and tariffs. It is no longer possible to maintain the large enterprises, rural households and rural village life by protecting inefficient producers from world competition. The participation of Russia in world economic systems, and the demands by Russian consumers for high quality food products makes it impossible to isolate Russian agriculture from world markets. The critical question then is whether it is possible to maintain viable peasant households, villages and an agricultural sector in a Russia that is increasingly becoming integrated into a world economic system?

Our study provides some hopeful answers to this question. There is sufficient human and social capital in rural areas in a number of economic regions to support a viable household-based agriculture. Our data shows that a substantial number of rural households already have begun the process of adapting to the demands of the market economy. Demands for protectionism and subsidies are not coming from peasant households, but rather from the large enterprises and the political figures at the local, regional and national levels that are the primary source of resistance to change. It is true that rural households tend to vote conservative in elections. This conservatism, however, stems largely from a distrust of reformers in Moscow who have, so far, not shown much interest in supporting the development of household enterprises.

In their actual *behavior,* rural Russian households have shown a considerable amount of entrepreneurial spirit as they have tried to cope with the problems of an emerging market economy. Even households with limited capital, increased their production and sales during the study. Peasant households have actually increased the proportion they contribute to overall Russian agricultural output, now accounting for the production of over two-thirds of the vegetables, half of the meat and over 45 percent of dairy products.

There are, however, two conditions that have been observed during the study that threaten both the viability of rural households and the viability of Russian villages. The first is that the maximum growth of household enterprises *within the constraints of current land tenure and credit arrangements* has been reached by some households. These households have reached a threshold that the contribution of additional human labor and social capital, or even basic forms of physical capital, will increase their productivity and sales. What is needed by these households are access to more land *and* credit to purchase the equipment and inputs necessary to work larger plots of land and care for larger numbers of animals.

We have not found any evidence to support the view that somehow Russian peasants have a cultural resistance to private farming or that they are unreasonably cautious about expanding their operations. Peasant households are reluctant to totally cut their ties to the *kolkhozy* or *TOO* simply because they are rightly fearful that such a complete break would cut them off from much needed services to support their household enterprises and the health and well-being of their families. These households are very rational in their approach to their family business operations. They are cautious, much like the German Catholic farmers in Sonya Salamon's (1985) study of the rural US, in that their primary goal is to preserve the household at all costs. This means eschewing risky decisions that would threaten the ability of the household to survive in a politically and economically unstable environment. Not surprisingly, those who were most likely to take the risk to disassociate themselves from the large enterprises and become officially registered private farmers (*fermery*) are urban dwellers who in most instances have not had much historic investment or attachment to a rural village. In contrast to the peasant

Supporting Sustainable Households and Communities 239

households, their primary risk is money and not a more diffuse connection with a "family way of life."

The primary means by which peasant households have expanded their agricultural productivity and sales has been to more efficiently utilize human, social, and physical capital. Households with more household labor, more extensive helping networks, and greater community involvement have been able to obtain more animals, more land, and more mechanical equipment. In turn, they have produced substantially more than have their neighbors and have sold more goods in the marketplace.

Although a significant number of rural Russian households have managed to improve their productivity, their potential for growth has been limited by the slow pace of institutional change at the village, regional, and national levels. Specifically, there has been very limited development of a market infrastructure, including land and credit markets, *geared toward the needs of households and not large enterprises* and a failure to develop formal-legal, member-owned and managed cooperatives. Institutional structures complement and increase the effectiveness of informal and formal exchange relations by providing safeguards and enforcement (North 1990). Stunted institutional growth in these two areas increases the likelihood that peasant households, *as rational economic actors,* will have short time horizons, minimal fixed capital and not take the risks necessary to expand their production.

The second condition that threatens the viability of rural households and villages *is the failure of different levels of government and the private sector to deal with the social service needs* that have emerged in the post-Soviet era. In an earlier work we dealt with this at some length (see O'Brien et al. 1998b). Among those households that have made more successful economic adaptations to the market economy, due largely to their higher levels of household capital, there is a very real basis of concern that they will not be able to meet their health, education and other social service needs. This fear is a major source of resistance to so-called reforms that would simply introduce land sales or other "marketplace" measures without dealing with the problem of meeting social service needs.

The problems associated with the lack of institutional development in the social service sphere have been even greater for those households that have not fared so well in the post-Soviet economy. This study has shown that there has been a very uneven adaptation of households to a market economy. This differentiation in both the material conditions and

240 Household Capital and the Agrarian Problem in Russia

psychological adjustment of households represents an emerging system of inequality in the countryside. The problem of "fairness", that results from the unequal abilities of different households to compete in a marketplace is a perennial problem in democratic societies. These democracies, as Schumpeter (1950) observed, have been able to weaken the disruptive effects of such inequalities by institutionalizing various kinds of incremental welfare state adjustments. Social security or welfare measures, instituted in other industrial democracies, have provided a "safety net" to either substitute or supplement the limited household capital of more vulnerable portions of the population.

Supporting Sustainable Households and Communities 241

Figure 12.1: Future Institutional Structure of Rural Life in Russia

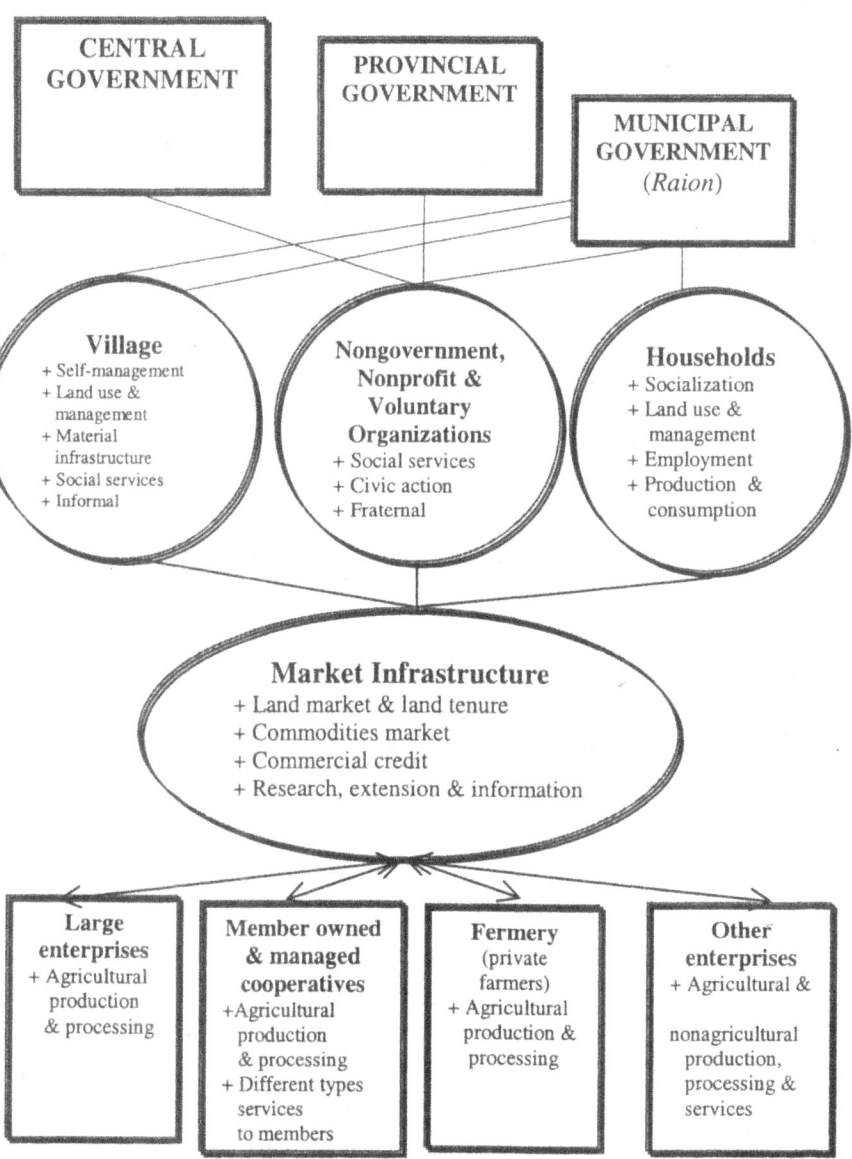

The collapse of the Soviet system of social welfare and the financial crisis facing the Russian central government, however, has meant that no new system of social welfare has replaced the one that has been abandoned. This poses a direct challenge to the viability of rural villages, in which over one-third of the population is elderly.

In order to preserve the viability of Russian rural households and villages and facilitate the movement of Russian agriculture into the international agricultural economy, it will be necessary to make some important institutional changes that go beyond the adjustments that have already been made in the immediate post-Soviet era.

The types of institutional changes that are necessary to create viable household farming enterprises and viable villages in a rural Russia that is a participant in a world agricultural system are illustrated in Figure 12.1. A useful point of comparison are Figures 1.1 and 1.2 in Chapter I that outline the institutional structure of rural life in the Soviet and immediate Post-Soviet periods. The figure presented above is a projection of what is possible to achieve in the next decade or so.

Figure 12.1 assumes that some type of large-scale capital intensive enterprises will continue to produce most of the grain, sugar beets and sunflower seed for oil in Russia. Most likely, these enterprises will be new types of cooperatives or some restructured variant of the current joint-stock companies. This would not be that different, for example, than the joint stock companies that currently produce most of the grain in the Czech Republic. Questions pertaining to the ownership of shares by current members of joint stock companies, the dividends received for such shares, and the transfer of ownership of shares, are being seriously discussed at the national and local levels. There is good reason for optimism that these issues can be resolved.

The more difficult task, however, is to maintain a viable economic niche for household producers, which means not only supporting their agricultural production, but also supporting other needs of the household and the village community upon which they depend for support.

The immediate post-Soviet period did produce several changes that, in the long run, are critical for the development of household enterprises. The weakening of Soviet era restrictions on household production and sales legitimized the role of the household as a producer and seller of agricultural products in a way that would have been unheard of even in the late 1980s. The reduction of the central government's authority has permitted the

Supporting Sustainable Households and Communities 243

beginning of the devolution of power to provincial and municipal governments, as well as to village administrations. Although all of these latter units of government are greatly under funded and, in many ways, still very ineffective, there is some evidence that they can find creative ways to improve conditions for the expansion of household enterprises. In this study, the Belgorod *Oblast* program to provide credit for expansion of household enterprises was shown to have produced substantial increases in household productivity and sales. There is every reason to believe that competition between provinces, *raions*, and villages will increase the development of creative programs to improve household opportunities for production and sales. Elsewhere we have shown that there is growing differentiation between economic regions, provinces and villages with respect to strategies they employ to deal with household enterprises (see O'Brien et al. 1998).

There are a number of issues that will require changes in the relationships between different levels of government and between governments, the private sector, and not for profit organizations. The most pervasive issue in this regard is an ongoing struggle over spheres of authority and influence. This is not played out so much in terms of different interests trying to gain influence but rather in terms of trying to avoid responsibility. Given the current economic crisis facing Russia, there is a tendency for all levels of government to avoid responsibility for paying the costs of social and material infrastructure.

In democratic states, the issue of spheres of influence of different levels of government and government versus the private and not for profit sectors is an ongoing concern. Thus, the lack of a resolution of this issue is not necessarily a sign of any fundamental problem in Russian political life. Moreover, as noted earlier, the loss of prerogatives by the Russian central government has resulted in much more activity from provincial and regional governments. In time, in our view, this will produce some creative solutions to institutional change.

Institutional structures that have the most direct bearing on the agrarian problem and sustainable households in rural Russia are those that support household enterprises. During the whole Soviet era, the monopoly of agriculture by the government meant that there were no formal commodity or commercial credit markets. Rather, the state subsidized large agricultural enterprises and was the exclusive purchaser of their raw agricultural commodities. The state also monopolized processing facilities and access to the state stores that sold these products to consumers. Research information was funneled directly into the large enterprises. Agronomists and veterinarians working for these enterprises were state employees.

In order to create a market infrastructure that is more suitable for household enterprises it will be necessary to create several new types of organizations at the village level. In addition to a land market, it will be necessary to create formal markets for commodities and credit. Also, it will be necessary to create a way of channeling research findings to meet the specific needs of small- and medium-scale producers. This means creating an infrastructure that will link research institutes and households. One possible model here might be a type of extension or outreach system.

The biggest challenge in institutional development is the creation of member owned and managed cooperatives. The cooperative movement has been an essential element in the development of family farming in the United States and other countries. There are, of course, a variety of models from different parts of the world that would meet the specific preferences of Russian rural households. At minimum, however, these cooperatives would have to deal with the specific problems of small producers in obtaining inputs (e.g., seeds and fertilizers) and purchasing various kinds of services.

The inter-governmental issue that has most bearing on the viability of household enterprises is the need to create some type of legitimized land tenure and use system that will create a land market. There are a number of ways in which land tenure and use is handled. The issue here is not ownership *per se*. A large proportion of the land used by Midwestern American farmers, for example, is rented. The critical issue is whether there exists a system in which households can *have trust* that the land they are planning to use to pay off debts for seed or fertilizer or equipment will be available to them for specified periods of time under specified conditions.

Another major concern with respect to the viability of rural households and villages is the extent to which self-government can be institutionalized in the village. During the Soviet era, local government, the *sel'soviet*, essentially was a clerical department in a large bureaucracy. In that system, the *kolkhoz* or *sovkhoz* was directly responsible for the production and delivery of social services, and building the material infrastructure of the village. Under these conditions, the chairpersons needed to consider more

Supporting Sustainable Households and Communities 245

than just the efficiency of their production enterprises. They were responsible for the social needs of all village residents as well.

In market economies, an economic enterprise would be involved primarily in production of goods and/or services. Various levels of government have oversight on the development of the material infrastructure, such as transportation, communication and electrification. Social services are primarily the responsibility of a mix of various levels of government, social welfare agencies, and nongovernmental charitable and religious organizations.

During the Soviet era, there was a lack of any significant civic voluntary organizations and charitable institutions. In the last five years, there has been considerable development of this aspect of civic life in Russia. For the most part, however, the development of civic institutions has occurred in urban rather than rural areas of Russia (Dershem and Patsiorkovski 1997). The difficulty of creating and maintaining non-governmental organizations in rural areas is due to the lack of community organizations, individual experience and material resources. Clearly, a variety of models of social service delivery exists. There is every reason to expect that if local people and government are involved in the community development process, social change will lead to a resolution of the agrarian problem and support sustainable households and villages in rural Russia.

APPENDIXES

248 Household Capital and the Agrarian Problem in Russia

Appendix 1: Chronology of Land Reform Laws in Russia

Информационные каналы Государственной Думы РФ, Совета Федерации РФ: ЗЕМЕЛЬНЫЙ КОДЕКС РОССИЙСКОЙ ФЕДЕРАЦИИ". М.: Интернет-сервер "АКДИ Экономика и жизнь", 1998 <http://www.akdi.ru/gd/proekt/GD05.HTM>.

1. Внесен Распоряжением Правительства РФ от 19.06.94 N 953-р, депутатами Государственной Думы, согласительной комиссией.
2. Официальные представители Правительства РФ – Председатель Роскомзема Н.В. Комов, Министр сельхозпрода России В.Н. Хлыстунов, Министр юстиции России Ю.Х. Калмыков.
3. Профильный комитет - Комитет по аграрным вопросам.
4. Профильный комитет Совета Федерации - Комитет по аграрной политике.
5. Принят Государственной Думой к рассмотрению (Постановление N 163-1 ГД) 8 июля 1994 года.
6. Принят Государственной Думой в первом чтении (Постановление N 993-1 ГД) 14 июля 1995 года.
7. Принят Государственной Думой во втором чтении (Постановление N 257-II ГД) 17 апреля 1996 года.
8. Принят Государственной Думой в третьем чтении (Постановление N 360-II ГД) 22 мая 1996 года.
9. Принят к рассмотрению Советом Федерации (Постановление N 225-СФ) 5 июня 1996 года.
10. Отклонен Советом Федерации (Постановление N 244-СФ) 26 июня 1996 года.
11. Попытка преодоления вето Совета Федерации 12 июля 1996 года.
12. Создание согласительной комиссии (Постановления N 244-СФ, 21-СФ, 543-II ГД) 26 июня 1996 года, 12 июля 1996 года, 23 января 1997 года.
13. Принят Государственной Думой в третьем чтении в редакции согласительной комиссии (Постановление N 1507-II ГД) 11 июня 1997 года.
14. Одобрен Советом Федерации (Постановление N 248-СФ) 3 июля 1997 года.
15. Отклонен Президентом Российской Федерации 21 июля 1997 года. (Президент: Земельный кодекс РФ содержит положения, которые противоречат Конституции РФ и Гражданскому кодексу РФ. При работе над Кодексом проигнорированы рекомендации ведущих ученых в области земельного и гражданского права. В целом Кодекс следует охарактеризовать как законодательный акт, не только грубо нарушающий конституционные права граждан и юридических лиц на землю, но и игнорирующий интересы государства в области использования и охраны земель. Указанные недостатки могут быть устранены только в процессе работы согласительной комиссии с участием ведущих ученых в области земельного и гражданскогоправа.)
16. Одобрен профильным комитетом в прежней редакции и внесен в Совет Государственной Думы 17 сентября 1997 года.
17. Принят Государственной Думой в третьем чтении в прежней редакции (Постановление N 1735-II ГД) 24 сентября 1997 года.
18. Принят к рассмотрению Советом Федерации 15 октября 1997 года.
19. Советом Федерации перенесено рассмотрение 28 января 1998 года.
20. Отклонен Советом Федерации (не преодолено вето)
21. (Постановление N 40-СФ) 18 февраля 1998 года.
22. Принят Государственной Думой в третьем чтении с учетом предложений Президента РФ (Постановление N 2405-II ГД) 22 апреля 1998 года.
23. Одобрен Советом Федерации 20 мая 1998 года.
24. Возвращен Президентом РФ без рассмотрения 3 июня 1998 года. (Президент: не могу согласиться с решением Государственной Думы, что кодекс принят с учетом предложений Президента РФ, поскольку внесенные в его текст изменения в ряде случаев не соответствуют предложениям, изложенным в письме Президента РФ от 21 июля1997 г.
25. Представленный на рассмотрение Государственной Думы Земельный кодекс, по существу, являлся новым законопроектом, в отношении которого должно быть соблюдено установленное частью 3 статьи 104 Конституции РФ требование о наличии заключения Правительства РФ. Отсутствие такого заключения не позволяет Президенту РФ считать Земельный кодекс принятым в соответствии с Конституцией РФ.

Appendixes 249

Appendix 2: Rural Population by Economic Regions and Major Territorial-Administrative Divisions in Russia from 1970 to 1997 (percent)

Regions	1970	1979	1989	1997
Russian Federation	37.7	30.7	26.5	26.9
North Region	**35.6**	**27.5**	**23.5**	**24.2**
Republic of Karelia	31.3	22.2	18.3	26.4
Republic of Komi	40.1	29.0	24.4	25.6
Arkhangelsk *Oblast*	34.3	27.9	26.6	26.2
Nenets autonomus area	46.2	40.4	38.2	40.6
Vologda *Oblast*	52.5	41.3	35.0	32.0
Murmansk *Oblast*	11.4	10.6	8.0	8.0
The North West Region	**20.1**	**15.4**	**13.4**	**13.3**
The City of Sankt-Peterburg	-	-	-	-
Leningrad *Oblast*	41.3	36.0	34.1	34.1
Novgorod *Oblast*	46.5	35.3	30.4	29.0
Pskov *Oblast*	57.3	44.7	36.9	34.3
The Central Region	**28.7**	**21.4**	**17.4**	**17.0**
Bryansk *Oblast*	52.6	41.1	32.7	31.3
Vladimir *Oblast*	29.0	24.9	20.8	19.7
Ivanovo *Oblast*	23.5	24.9	20.8	17.7
Kaluga *Oblast*	48.1	37.9	31.1	25.8
Kostroma *Oblast*	46.6	36.2	31.4	33.5
The City of Moscow	-	-	-	-
Moscow *Oblast*	31.3	25.4	20.7	20.3
Oryel *Oblast*	61.1	44.8	37.7	36.9
Ryazan *Oblast*	52.9	41.7	34.2	31.7
Smolensk *Oblast*	52.1	40.1	32.0	30.1
Tver *Oblast*	43.2	33.4	28.5	27.1
Tula *Oblast*	28.7	21.9	19.0	18.6
Yaroslavl *Oblast*	29.7	22.2	18.4	19.5
Volga-Vyatka Region	**47.2**	**37.8**	**31.1**	**28.8**
Republic of Mariy El	59.1	46.8	38.8	37.8
Republic of Mordovia	63.8	53.0	43.5	40.9
Chuvash Republic	64.3	54.0	42.1	39.2
Kirov *Oblast*	45.3	35.7	30.1	29.4
Nizhny Novgorod *Oblast*	35.7	27.4	30.0	21.9
Black-earth Region	**59.8**	**47.9**	**39.7**	**38.1**
Belgorod *Oblast*	64.8	47.3	36.9	34.9
Voronezh *Oblast*	54.5	45.9	39.1	38.1
Kursk *Oblast*	67.0	52.3	42.1	39.5
Lipetsk *Oblast*	55.7	43.7	37.4	36.1
Tambov *Oblast*	60.6	51.1	43.1	41.9

Regions	1970	1979	1989	1997
Volga Region	**40.6**	**32.2**	**26.7**	**26.9**
Republik of Kalmykia	65.7	59.5	54.0	60.8
Republic of Tatarstan	48.5	36.8	26.9	26.4
Astrakhan *Oblast*	39.4	32.6	31.8	33.2
Volgograd *Oblast*	34.4	28.5	24.1	25.6
Penza *Oblast*	55.8	45.1	38.1	35.7
Samara *Oblast*	28.4	21.9	19.2	19.5
Saratov *Oblast*	34.9	28.9	26.7	26.7
Ulyanovsk *Oblast*	47.7	36.6	28.6	27.1
North-Caucasus Region	**50.2**	**45.1**	**42.7**	**44.5**
Republic of Adygeya	60.5	52.1	47.8	46.1
Republic of Dagestan	64.7	60.7	56.4	58.2
Republic of Ingushetia	58.3	57.5	58.5	58.2
Chechen Republic	58.3	57.5	58.5	65.3
Kabardian-Balkar Republic	52.4	41.8	38.8	42.4
Karachaev-Chercassian Republic	67.3	56.5	51.2	55.6
Republic of North Ossetia-Alania	35.5	32.2	31.2	31.0
Krasnodar *Kray*	52.3	47.8	45.5	45.9
Stavropol *Kray*	55.8	49.1	45.6	45.9
Rostov *Oblast*	36.8	31.2	28.7	32.2
The Ural Region	**35.4**	**29.1**	**25.3**	**25.6**
Republic of Bashkortostan	54.7	43.2	36.2	35.4
Udmurt Republic	42.9	34.7	30.3	30.3
Kurgan *Oblast*	57.2	49.4	45.3	45.0
Orenburg *Oblast*	46.9	39.7	35.0	35.9
Perm *Oblast*	32.5	26.1	22.7	23.7
Komi-Permyatsky autonomus area	81.1	75.8	70.0	69.2
Sverdlovsk *Oblast*	19.3	15.0	12.9	12.6
Chelyabinsk *Oblast*	22.1	18.8	17.5	18.7
West Siberian Region	**38.6**	**32.3**	**27.2**	**29.0**
Republic of Altai	76.2	72.1	72.9	78.0
Altai *Kray*	52.5	50.0	41.8	47.5
Kemerovo *Oblast*	17.7	13.8	12.6	13.2
Novosibirsk *Oblast*	34.6	28.5	25.3	26.1
Omsk *Oblast*	44.7	37.2	32.2	32.6
Tomsk *Oblast*	40.7	34.5	31.0	33.9
Tyumen *Oblast*	50.9	39.1	24.0	24.0
Khanty-Mansi autonomus area	37.3	21.6	9.0	9.2
Yamalo-Nenets autonomus area	57.5	49.4	22.0	17.2
East Siberian Region	**38.3**	**31.3**	**28.1**	**28.6**
Republic of Buryatia	55.1	43.1	38.5	40.3

Regions	1970	1979	1989	1997
Republic of Tuva	62.3	57.7	52.8	53.1
Republic of Khakassia	40.4	30.8	27.5	29.3
Krasnoyarsk *Kray*	37.8	30.8	27.1	25.9
Taimyr autonomus area	39.5	34.0	32.7	34.2
Evenki autonomus area	69.2	62.5	70.8	70.4
Irkutsk *Oblast*	27.7	22.5	19.5	20.5
Ust-Ordyn Buryat autonomus area	83.1	81.2	81.6	100.0
Chita *Oblast*	42.5	37.0	34.8	37.1
Aginsky Buryat autonomus area	78.8	73.9	67.5	67.7
Far East Region	**28.5**	**25.5**	**24.2**	**24.3**
Republic of Sakha (Yakutia)	43.5	38.7	33.3	35.7
Jewish autonomus *Oblast*	31.4	31.5	33.8	32.5
Chukchi autonomus area	30.7	30.1	27.4	30.2
Primorsky *Kray*	27.1	24.2	22.5	21.9
Khabarovsk *Kray*	20.9	19.2	20.0	19.5
Amur *Oblast*	38.2	35.0	32.3	35.2
Kamchatka *Oblast*	24.0	17.2	18.7	19.2
Koryak autonomus area	67.7	58.8	61.5	75.1
Magadan *Oblast*	23.0	18.6	15.5	10.0
Sachalin *Oblast*	21.5	17.6	16.2	13.9
Kaliningrad *Oblast*	26.8	23.5	20.9	22.2

Source: *Demograficheskii ezhegodnik Rossii*. Official Publication. Moscow: Goskomstat of Russia. 1997. pp.30- 32.

Appendix 3: Demographic Types of Households in the Panel Study by Village from 1995 to 1997 (n=463)

	1995			1996			1997		
	Laton n=157	Veng n=156	Sviatt n=150	Laton n=157	Veng n=156	Sviatt n=150	Laton n=157	Veng n=156	Sviatt n=150
1. Single person	15.9	21.8	27.3	14.6	20.5	27.3	15.9	21.2	28.7
2. Retired couple	12.1	9.0	8.7	12.7	9.0	10.0	14.0	9.6	10.0
3. Employed couple	7.0	5.8	12.0	4.5	5.8	11.3	3.8	5.8	11.3
4. Employed couple with children	36.9	28.8	26.0	36.3	29.5	28.0	36.3	30.1	24.0
5. Employed couple with children & other adults	10.8	14.7	4.0	14.0	15.4	2.7	15.3	17.3	5.3
6. Single parents	5.1	1.3	1.3	8.3	1.3	2.0	7.0	.6	2.0
7. Other	12.1	18.6	20.7	9.6	18.6	18.7	7.6	15.4	18.7
Total	100.0	100.0	100.0	100.0	100.0	100.0	100.0	100.0	100.0

Appendix 4: Distribution of the Rural Population in Russia by Economic Regions and Major Territorial-Administrative Division from 1970 to 1997 (percent)

Regions	1970	1979	1989	1997
The Russian Federation	100.0	100.0	100.0	100.0
The North Region	3.8	3.7	3.7	3.6
Republic of Karelia	0.5	0.4	0.4	0.5
Republic of Komi	0.8	0.8	0.8	0.8
Arkhangelsk *Oblast*	0.9	1.0	1.1	1.0
Nenets autonomus area	-	-	-	-
Vologda *Oblast*	1.4	1.3	1.2	1.1
Murmansk *Oblast*	0.2	0.2	0.2	0.2
The North West Region	2.9	2.8	2.8	2.7
The City of Sankt-Peterburg	-	-	-	-
Leningrad *Oblast*	1.1	1.3	1.5	1.5
Novgorod *Oblast*	0.7	0.6	0.6	0.5
Pskov *Oblast*	1.1	0.9	0.9	0.7
The Central Region	16.3	14.8	13.6	12.7
Bryansk *Oblast*	1.7	1.5	1.2	1.1
Vladimir *Oblast*	1.0	0.9	0.9	0.8
Ivanovo *Oblast*	0.6	0.6	0.6	0.5
Kaluga *Oblast*	0.9	0.8	0.8	0.7
Kostroma *Oblast*	0.9	0.7	0.7	0.6
The City of Moscow	-	-	-	-
Moscow *Oblast*	3.7	3.7	3.5	3.4
Oryel *Oblast*	1.1	1.0	0.9	0.9
Ryazan *Oblast*	1.6	1.3	1.2	1.1
Smolensk *Oblast*	1.2	1.1	1.0	0.9
Tver *Oblast*	1.6	1.4	1.3	1.1
Tula *Oblast*	1.1	1.0	0.9	0.9
Yaroslavl *Oblast*	0.9	0.8	0.8	0.7
Volga-Vyatka Region	8.1	7.5	6.8	6.3
Republic of Mariy El	0.9	0.8	0.7	0.7
Republic of Mordovia	1.3	1.2	1.1	1.0
Chuvash Republic	1.6	1.5	1.5	1.4
Kirov *Oblast*	1.6	1.4	1.3	1.2
Nizhny Novgorod *Oblast*	2.7	2.6	2.2	2.0
Black-earthRegion	9.7	8.8	7.9	7.5
Belgorod *Oblast*	1.5	1.5	1.2	1.2

254 Household Capital and the Agrarian Problem in Russia

Regions	1970	1979	1989	1997
Voronezh Oblast	2.6	2.6	2.5	2.4
Kursk Oblast	2.2	1.8	1.5	1.4
Lipetsk Oblast	1.2	1.2	1.2	1.1
Tambov Oblast	2.2	1.7	1.5	1.4
Volga Region	**12.0**	**11.9**	**11.2**	**11.5**
Republik of Kalmykia	0.4	0.4	0.5	0.5
Republic of Tatarstan	3.1	3.1	2.9	2.7
Astrakhan Oblast	0.7	0.8	0.9	0.8
Volgograd Oblast	1.6	1.6	1.8	1.7
Penza Oblast	1.7	1.6	1.6	1.5
Samara Oblast	1.6	1.5	1.5	1.4
Saratov Oblast	1.7	1.7	1.8	1.7
Ulyanovsk Oblast	1.2	1.2	1.2	1.2
North-Caucasus Region	**14.6**	**16.6**	**18.3**	**19.9**
Republic of Adygeya	0.5	0.5	0.5	0.5
Republic of Dagestan	2.2	2.4	2.6	2.9
Republic of Ingushetia	0.3	0.4	0.5	0.5
Chechen Republic	1.0	1.1	1.4	1.4
Kabardian-Balkar Republic	0.6	0.7	0.8	0.9
Karachaev-Chercassian Republic	0.5	0.5	0.5	0.6
Republic of North Ossetia-Alania	0.4	0.4	0.4	0.5
Krasnodar Kray	4.3	5.0	5.5	5.9
Stavropol Kray	2.2	2.4	2.8	3.1
Rostov Oblast	2.6	3.2	3.3	3.6
The Ural Region	**13.7**	**13.4**	**13.2**	**13.2**
Republic of Bashkortostan	4.0	3.9	3.7	3.7
Udmurt Republic	1.1	1.1	1.1	1.1
Kurgan Oblast	1.2	1.2	1.2	1.2
Orenburg Oblast	2.1	2.0	1.9	2.1
Perm Oblast	2.2	2.1	2.1	1.9
Komi-Permyatsky autonomus area	0.2	0.2	0.2	0.2
Sverdlovsk Oblast	1.6	1.6	1.5	1.3
Chelyabinsk Oblast	1.3	1.3	1.5	1.7
West Siberian Region	**9.5**	**9.9**	**10.5**	**11.0**
Republic of Altai	0.3	0.3	0.3	0.4
Altai Kray	2.7	2.7	2.8	3.1
Kemerovo Oblast	1.0	0.9	1.0	1.0

Regions	1970	1979	1989	1997
Novosibirsk *Oblast*	1.7	1.7	1.8	1.8
Omsk *Oblast*	1.6	1.7	1.8	1.8
Tomsk *Oblast*	0.6	0.7	0.8	0.8
Tyumen *Oblast*	1.3	1.6	1.8	1.9
Khanty-Mansi autonomus area	0.2	0.2	0.1	0.1
Yamalo-Nenets autonomus area	0.1	0.1	0.1	0.1
East Siberian Region	**5.8**	**6.1**	**6.6**	**6.6**
Republic of Buryatia	0.8	0.8	0.9	0.9
Republic of Tuva	0.2	0.3	0.4	0.4
Republic of Khakassia	0.3	0.3	0.4	0.4
Krasnoyarsk *Kray*	2.0	1.9	1.9	1.9
Taimyr autonomus area	-	-	-	-
Evenki autonomus area	-	-	-	-
Irkutsk *Oblast*	1.2	1.3	1.4	1.4
Ust-Ordyn Buryat autonomus area	0.2	0.3	0.4	0.4
Chita *Oblast*	1.0	1.1	1.1	1.1
Aginsky Buryat autonomus area	0.1	0.1	0.1	0.1
Far East	**3.4**	**4.1**	**4.9**	**4.5**
Republic of Sacha	0.6	0.8	0.9	0.9
Jewish autonomus *Oblast*	0.1	0.1	0.1	0.1
Chukchi autonomus area	0.1	0.1	0.1	0.1
Primorsky *Kray*	1.0	1.2	1.4	1.3
Khabarovsk *Kray*	0.5	0.6	0.9	0.8
Amur *Oblast*	0.6	0.8	0.9	0.9
Kamchatka *Oblast*	0.1	0.1	0.2	0.2
Koryak autonomus area	-	-	-	-
Magadan *Oblast*	0.1	0.1	0.1	-
Sachalin *Oblast*	0.3	0.3	0.3	0.2
Kaliningrad *Oblast*	0.4	0.4	0.5	0.5

Source: *Demograficheskii ezhegodnik Rossii*. Official Publication. Moscow: Goskomstat of Russia. 1997. pp.30- 32.

256 Household Capital and the Agrarian Problem in Russia

Appendix 5: Number and Size of Rural Households by Economic Regions and Major Territorial-Administrative Divisions in Russia in 1994

Economic Regions & Major Territorial-Administrative Divisions	Number of House-holds (in 1000)	From 1 person	From 2 persons	From 3 persons	From 4 persons	From 5 & more persons	Average size of House-holds
1	2	3	4	5	6	7	8
The Russian Federation	**13850.6**	**220**	**268**	**181**	**190**	**141**	**2.85**
The North Region	**541.0**	**241**	**265**	**192**	**197**	**105**	**2.71**
Republic of Karelia	79.2	250	290	197	180	83	2.59
Republic of Komi	109.5	217	248	199	210	126	2.84
Arkhangelsk *Oblast*	148.3	237	253	182	205	123	2.79
Nenets autonomous area	6.0	210	204	201	216	169	3.06
Vologda *Oblast*	175.0	267	287	178	178	90	2.58
Murmansk *Oblast*	28.5	175	202	292	260	71	2.88
The North West Region	**439.4**	**302**	**286**	**175**	**158**	**79**	**2.46**
The City of Sankt-Peterburg	-	-	-	-	-	-	-
Leningrad *Oblast*	212.7	242	267	205	191	95	2.67
Novgorod *Oblast*	93.6	341	286	158	147	68	2.35
Pskov *Oblast*	133.1	360	314	145	118	63	2.25
The Central Region	**2111.0**	**303**	**294**	**170**	**153**	**80**	**2.45**
Bryansk *Oblast*	190.6	311	300	152	151	86	2.44
Vladimir *Oblast*	136.5	324	287	161	154	74	2.40
Ivanovo *Oblast*	91.9	270	312	186	171	61	2.47
Kaluga *Oblast*	119.5	324	287	167	136	86	2.41
Kostroma *Oblast*	110.4	284	294	164	177	81	2.51
The City of Moscow	-	-	-	-	-	-	-
Moscow *Oblast*	513.3	255	264	205	184	92	2.64
Oryel *Oblast*	130.7	263	299	174	162	102	2.59

Continued

Appendixes 257

1	2	3	4	5	6	7	8
Ryazan *Oblast*	195.9	344	321	157	122	56	2.25
Smolensk *Oblast*	158.7	337	316	149	129	69	2.31
Tver *Oblast*	200.9	345	315	144	128	68	2.29
Tula *Oblast*	144.1	336	306	151	128	79	2.35
Yaroslavl *Oblast*	118.5	320	294	168	149	69	2.38
Volga-Vyatka Region	**945.4**	**255**	**266**	**172**	**187**	**120**	**2.71**
Republic of Mariy El	93.6	180	227	175	228	190	3.11
Republic of Mordovia	158.3	295	275	160	176	94	2.54
Chuvash Republic	185.5	250	242	150	160	198	2.93
Kirov *Oblast*	181.7	228	270	188	208	106	2.74
Nizhny Novgorod *Oblast*	326.3	271	281	180	186	82	2.55
Black-earth Region	**1217.3**	**296**	**301**	**162**	**149**	**92**	**2.49**
Belgorod *Oblast*	198.2	282	298	155	150	115	2.57
Voronezh *Oblast*	391.2	300	304	161	150	85	2.46
Kursk *Oblast*	218.0	306	313	146	133	102	2.47
Lipetsk *Oblast*	184.3	295	286	172	161	86	2.49
Tambov *Oblast*	226.6	296	295	174	154	81	2.46
Volga Region	**1612.1**	**215**	**274**	**185**	**199**	**127**	**2.82**
Republik of Kalmykia	58.1	105	205	195	255	240	3.46
Republic of Tatarstan	332.2	205	254	174	195	172	2.98
Astrakhan *Oblast*	105.7	156	230	184	242	188	3.18
Volgograd *Oblast*	241.1	196	282	187	213	122	2.85
Penza *Oblast*	222.7	260	304	184	172	80	2.55
Samara *Oblast*	238.2	222	279	202	203	94	2.70
Saratov *Oblast*	258.6	228	281	189	190	112	2.73
Ulyanovsk *Oblast*	155.5	244	302	169	188	97	2.63
North-Caucasus Region	**2251.1**	**171**	**233**	**175**	**199**	**222**	**3.27**

Continued

1	2	3	4	5	6	7	8
Republic of Adygeya	67.9	193	258	172	187	190	3.05
Republic of Dagestan	271.0	136	134	130	165	435	4.14
Republic of Ingushetia	21.5	12	36	46	83	823	7.13
Chechen Republic	79.5	12	36	46	83	823	7.13
Kabardian-Balkar Republic	68.7	108	139	138	184	431	4.16
Karachaev-Chercassian Republic	62.2	131	196	151	195	327	3.65
Republic of North Ossetia-Alania	51.5	143	185	154	174	344	3.77
Krasnodar *Kray*	771.4	199	262	187	201	151	2.94
Stavropol *Kray*	387.8	173	250	176	213	188	3.12
Rostov *Oblast*	470.6	167	268	200	218	147	2.99
The Ural Region	**1753.7**	**187**	**262**	**183**	**209**	**159**	**2.98**
Republic of Bashkortostan	456.6	158	243	193	214	192	3.14
Udmurt Republic	156.8	175	233	173	226	193	3.13
Kurgan *Oblast*	183.7	225	288	180	187	120	2.74
Orenburg *Oblast*	260.1	164	257	198	225	156	3.03
Perm *Oblast*	203.7	200	258	177	2-1	164	2.96
Komi-Permyatsky auto.area	37.1	202	256	170	184	188	3.02
Sverdlovsk *Oblast*	220.9	248	290	163	180	119	2.70
Chelyabinsk *Oblast*	234.8	179	273	183	226	139	2.93
West Siberian Region	**1481.8**	**167**	**277**	**199**	**214**	**143**	**2.96**
Republic of Altai	47.1	159	238	201	229	173	3.11
Altai *Kray*	439.5	155	295	207	217	126	2.92
Kemerovo *Oblast*	140.2	187	284	182	217	130	2.89
Novosibirsk *Oblast*	248.9	175	284	199	208	134	2.90
Omsk *Oblast*	228.7	161	270	178	208	183	3.10
Tomsk *Oblast*	129.5	178	268	208	211	135	2.93
Tyumen *Oblast*	187.1	169	255	208	224	144	3.00

Continued

1	2	3	4	5	6	7	8
Khanty-Mansi autonomus area	35.2	126	227	260	258	129	3.10
Yamalo-Nenets autonomus area	25.6	107	209	292	286	106	3.16
East Siberian Region	**828.7**	**154**	**247**	**191**	**209**	**199**	**3.18**
Republic of Buryatia	128.8	148	229	180	202	241	3.34
Repubblic of Tuva	39.3	81	146	186	201	386	4.06
Republic of Khakassia	52.9	166	262	182	211	179	3.07
Krasnoyarsk *Kray*	274.8	171	283	203	205	138	2.93
Taimyr autonomus area	5.0	177	236	170	193	224	3.19
Evenki autonomus area	5.5	146	191	278	246	139	3.11
Irkutsk *Oblast*	137.8	159	243	185	209	204	3.18
Ust-Ordyn Buryat auton. area	42.4	140	238	182	188	252	3.38
Chita *Oblast*	128.3	133	228	193	226	220	3.33
Aginsky Buryat autonomus area	13.9	98	162	179	215	346	3.90
Far East Region	**601.9**	**152**	**232**	**220**	**223**	**173**	**3.15**
Republic of Sakha (Yakutia)	93.5	90	158	183	223	346	3.89
Jewish autonomus *Oblast*	22.4	129	256	214	217	184	3.20
Chukchi autonomus area	12.0	172	262	293	206	67	2.76
Primorsky *Kray*	172.6	171	265	218	211	135	2.95
Khabarovsk *Kray*	100.7	143	224	215	253	165	3.16
Amur *Oblast*	122.7	173	250	228	213	136	2.95
Kamchatka *Oblast*	20.0	160	209	278	247	106	2.96
Koryak autonomus area	7.3	129	229	278	268	96	3.00
Magadan *Oblast*	15.8	224	224	272	181	99	2.73
Sachalin *Oblast*	34.9	159	238	237	250	116	2.98
Kaliningrad *Oblast*	67.2	187	269	195	200	149	2.96

Sources: *Rossiiski Statisticheski Ezhegodnik 1996*. Moscow: Logos. pp. 718-720; *Tipy I sostav Domokhosaistv v Rossii po mikroperepisi 1994 goda*. Moscow: Goscomstat Rossii. 1995. pp 23-32.

Appendix 6: Age Structure of the Rural Population in Economic Regions and Major Territorial-Administrative Divisions in 1997

Economic Regions	Total	Rural Population by Age Groups									
		7 & less	8-11	12-14	15-16	17-65	66-70	71-74	75-79	80 & more
The Russian Federation	39789817	4291923	2992946	2191762	131492	24234276	2356062	1172641	776902	1111722
The North Region	1416495	136887	110695	82256	47916	884499	79073	41055	25606	29992
Republic of Karelia	207761	18288	15215	11509	6825	133396	11245	5104	3103	3076
Republic of Komi	297812	28533	24411	18552	10609	190241	12225	5946	3384	3911
Arkhangelsk Oblast	394491	38131	31206	23167	13614	239374	21722	11642	7290	8345
Nenets auton. Area	18284	2419	1633	1179	678	11037	612	295	181	250
Vologda Oblast	430943	39419	30959	22469	13072	251327	31643	17336	11148	13570
Murmansk Oblast	85488	6897	7271	5380	3118	59124	1626	732	500	840
The North West Region	1072232	80597	69720	51367	31063	661517	74136	42119	26734	34979
The City of St. Petersburg										
Leningrad Oblast	569620	41173	38736	28884	17920	367444	33144	17531	10879	13909
Novgorod Oblast	216046	17612	14250	10333	5952	126606	17009	9401	6382	8501
Pskov Oblast	286566	21812	16734	12150	7191	167467	23983	15187	9473	12569
The Central Region	5056796	409205	315451	229332	137790	3048448	380182	217364	133686	185938
Bryansk Oblast	463996	43333	29018	20433	11542	260131	43102	23489	13824	19124
Vladimir Oblast	325660	26567	21092	15702	9646	195991	23876	13538	8237	11611
Ivanovo Oblast	222860	18000	14423	11026	6771	136886	15427	8326	4982	7019

Appendixes 261

Economic Regions	Total	7 & less	8-11	12-14	15-16	17-65	66-70	71-79	75-79	80 & more
Kaluga *Oblast*	284092	24188	18216	12923	7909	171832	20922	12347	6896	8859
Kostroma *Oblast*	266372	23413	18517	13514	7881	159523	18533	10469	6653	7869
City of Moscow	-	-	-	-	-	-	-	-	-	-
Moscow *Oblast*	1303296	96640	79931	59405	36862	844807	79585	43226	26246	36594
Oryel *Oblast*	339090	30196	20719	14384	8487	198678	28931	15669	8896	13130
Ryazan *Oblast*	428445	32132	23866	16937	10184	249201	35232	22105	15256	23532
Smolensk *Oblast*	353138	28848	22551	15954	9769	203294	29002	17635	11084	15001
Tver *Oblast*	448668	36587	28649	20716	11987	258271	36834	21869	14038	19717
Tula *Oblast*	336036	26474	20460	14934	8843	199146	28228	16048	9634	12269
Yaroslavl *Oblast*	285143	22827	18009	13404	7909	170688	20510	12643	7940	11213
Volga-Vyatka Region	**2530276**	**235247**	**175100**	**125398**	**75038**	**1492029**	**171232**	**96316**	**67327**	**92589**
Republic of Mariy El	289952	31653	23845	16561	9836	173085	16229	7965	5065	5713
Republic of Mordovia	392225	33352	24487	17476	10693	232486	28326	17087	11595	16723
Kirov *Oblast*	481220	45201	34856	26362	15260	286690	29816	16300	11895	14840
Nizhny Novgorod *Oblast*	828899	67471	52455	37934	22085	486083	63229	36642	25358	37642
Black-earth Region	**3019999**	**311726**	**180094**	**127990**	**76615**	**1777234**	**249761**	**129065**	**84796**	**142718**
Belgorod *Oblast*	519394	45453	31277	21402	12635	296999	45406	23693	15779	26750
Voronezh *Oblast*	959307	78499	57028	41371	24971	568270	79952	37587	24938	46691

262 Household Capital and the Agrarian Problem in Russia

Economic Regions	Total	Rural Population by Age Groups								
		7 & less	8-11	12-14	15-16	17-65	66-70	71-79	75-79	80 & more
Kursk *Oblast*	532562	43571	30607	21566	12709	309157	49140	26333	15775	23704
Lipetsk *Oblast*	455910	98119	28083	20070	11922	272046	34228	18824	12443	20175
Tambov *Oblast*	552826	46084	33099	23581	14378	330762	41035	22628	15861	25398
Volga Region	**4572637**	**471771**	**325757**	**232340**	**139537**	**2744127**	**281730**	**134988**	**92753**	**149634**
Republik of Kalmykia	193056	25906	18655	13488	7632	113548	6678	2832	1711	2606
Republic of Tatarstan	999776	111634	69004	45221	27635	581221	68741	33480	23946	38894
Astrakhan *Oblast*	340669	38972	27392	19704	11636	209080	16242	7076	4001	6566
Volgograd *Oblast*	698234	72813	49239	36403	21448	423220	41331	19016	12029	22735
Penza *Oblast*	559832	47038	35348	25605	15821	335251	39853	22909	15131	22876
Samara *Oblast*	643014	62121	45940	33196	19838	395707	38491	16151	12253	19317
Saratov *Oblast*	733610	75121	52726	38939	23520	445371	42073	20106	13917	21837
Ulyanovsk *Oblast*	404446	38166	27453	19784	12007	240729	28321	13418	9765	14803
North-Caucasus Region	7876219	983783	601191	435949	264258	4705894	396406	179697	122998	186043
Republic of Adygeya	207318	21581	14604	10552	6552	125428	12925	5657	3553	6466
Republic of Dagestan	1213627	230999	111642	76163	46920	659759	39977	16983	13024	18160
Republic of Ingushetia	179522	29873	16135	11006	6791	105564	4007	1679	1570	2897
Chechen Republic	529533	88433	51950	36361	22251	298348	14075	5446	4676	7993
Kabardian-Balkar Republic	333743	52614	28576	20093	12004	192515	13034	5029	3901	5977
Karachaev-Cher-cassian Republic	241569	31413	18848	13550	8099	144167	11816	4701	3322	5653

Appendixes 263

Economic Regions	Total	Rural Population by Age Groups									
		7 & less	8-11	12-14	15-16	17-65	66-70	71-79	75-79	80 & more
Republic of North Ossetia-Alania	206769	23188	15413	11421	6891	125199	11198	4817	3604	5038
Krasnodar *Kray*	2316541	226835	155678	117905	72082	1435291	134508	66063	44908	63271
Stavropol *Kray*	1226211	135489	89824	66614	40047	748232	66691	30179	19736	29399
Rostov *Oblast*	1421386	143358	98521	72284	42621	871391	88175	39143	24704	41189
The Ural Region	5224007	601001	429171	310049	182820	3153151	302097	133788	90795	127755
Republic of Bashkortostan	1450146	180966	118556	82696	49557	832515	86284	35728	24887	38957
Udmurt Republic	495075	55480	41610	30326	17510	296351	24898	11377	7972	9551
Kurgan *Oblast*	498996	50282	37858	28237	16715	298903	29917	13711	9269	14104
Orenburg *Oblast*	803111	93783	65177	45952	27603	477308	44484	17758	12058	18988
Perm *Oblast*	705869	77991	58034	42029	23952	419526	38560	18644	12330	14803
Sverdlovsk *Oblast*	585134	57180	44388	33412	19836	353027	34819	16921	11467	14084
Chelyabinsk *Oblast*	685676	72813	54572	40962	24102	413851	36853	16473	10829	15221
West Siberian Region	4397544	477371	362989	283594	170235	2808890	231504	108938	72439	91204
Republic of Altai	152920	20870	14381	10172	5954	88734	6250	2687	1669	2203
Altai *Kray*	1277844	119270	91163	73856	45420	784342	73881	35936	23391	30585
Kemerovo *Oblast*	403642	40915	31975	24705	14855	239749	23354	11490	7725	8874
Novosibirsk *Oblast*	720263	71743	54282	43877	26269	435482	41062	18984	12357	16207
Omsk *Oblast*	714045	82275	59279	46669	28396	420512	35936	16013	11170	13795

264 *Household Capital and the Agrarian Problem in Russia*

Economic Regions	Total	Rural Population by Age Groups								
		7 & less	8-11	12-14	15-16	17-65	66-70	71-79	75-79	80 & more
Tomsk *Oblast*	364765	35778	29640	23567	13761	226169	16833	8267	5259	5491
Tyumen *Oblast*	764065	80614	63524	47266	27738	477495	30638	14162	9883	12745
Khanty-Mansi autonomus area	123283	13796	10861	7981	4519	81235	2527	942	617	805
Yamalo-Nenets autonomus area	86337	12110	7884	5501	3323	55172	1023	457	368	499
East Siberian Region	2610883	350943	252692	186520	112787	1677705	122409	57090	38045	45387
Republic of Buryatia	420119	50881	39893	29676	17859	245378	16636	7836	5387	6573
Republic of Tuva	165356	29938	18292	12813	7773	89716	3181	1408	1107	1128
Krasnoyarsk *Kray*	804465	85051	63372	47823	28905	488168	42194	19977	13323	15652
Taimyr autonomus area	16444	2343	1538	1000	670	10375	215	113	117	73
Evenki autonomus area	14225	1751	1384	957	575	8921	288	121	97	131
Irkutsk *Oblast*	570234	70518	50892	37089	22429	335368	25750	11913	7733	8542
Ust-Ordyn Buryat autonomus area	142580	21194	14094	9685	5690	79080	5871	2789	1677	2500
Chita *Oblast*	479040	61140	43016	32492	19705	289850	18023	8199	5758	6857
Aginsky Buryat autonomus area	53446	8373	5676	4099	2545	29112	1586	634	534	887
Far East Region	1803259	215385	154414	115781	69391	1146942	57440	27293	19080	22223
Republic of Sakha (Yakutia)	364660	60628	36321	25347	14179	209218	9155	3530	2998	3284
Jewish autonomus *Oblast*	67699	8783	6307	4640	2703	40618	2053	986	681	928
Chukchi autonomus area	27380	3411	2576	1900	1152	17474	430	178	149	110

Economic Regions	Total	Rural Population by Age Groups									
		7 & less	8-11	12-14	15-16	17-65	66-70	71-74	75-79	80 & more
Primorsky *Kray*	493407	51679	38768	29315	18017	313144	18319	9508	6728	7929
Khabarovsk *Kray*	297533	31802	24508	18513	11387	192140	8851	4289	2864	3179
Amur *Oblast*	356675	37552	28764	22402	13452	225712	13337	6413	4109	4934
Kamchatka *Oblast*	78162	8036	6173	4891	2981	52817	1522	654	501	587
Koryak autonomus area	24690	2819	2039	1591	898	16319	453	221	181	169
Magadan *Oblast*	28808	2203	2040	1872	1267	19897	741	313	155	320
Sachalin *Oblast*	88935	8472	6918	5310	3355	59603	2579	1201	714	783
Kaliningrad *Oblast*	209470	21207	15672	11186	6642	133840	10092	4928	2643	3260

Source: *Chislennost' naselenia Rossiiskoi Federatsii po polu i vozrastu na 1 January 1997.* Moscow: Goskomstat Rossii.1997. pp. 46-246.

266 *Household Capital and the Agrarian Problem in Russia*

Appendix 7: Mean Percentage of Rented Land by Redundant Ties

Redundant Ties	1995	1996	1997
0-8	0.00 (n=132)	0.00 (n=113)	0.00 (n=98)
9-10	0.00 (n=105)	0.00 (n=81)	0.00 (n=66)
11-13	0.13 (n=109)	0.12 (n=108)	0.13 (n=112)
14-16	0.00 (n=117)	0.00 (n=161)	0.15 (n=187)
Total Sample (N=463)	0.00	0.00	0.12

Rent by redundant ties: 95-$F(4)=3.46$, $p<.05$, Scheffe, 1 x 3, $p<.05$; 96-$F(4) =2.32$, $p=n.s.(.074)$; 97-$F(4)=4.01$, $p<.01$, Scheffe, 1 x 4, $p<.01$

Appendix 8: Mean Number of Total Agricultural Tools (excluding tractors) by Number of Redundant Ties

Redundant Ties	1995	1996	1997
0-8	0.11 (n=132)	0.18 (n=113)	0.15 (n=98)
9-10	0.00 (n=105)	0.19 (n=81)	0.27 (n=66)
11-13	0.11 (n=109)	0.26 (n=108)	0.45 (n=112)
14-16	0.22 (n=117)	0.17 (n=161)	0.21 (n=187)
Total Sample (N=463)	0.14 (N=463)	0.19 (N=463)	0.27 (N=463)

1995-$F(3)=1.54$, $p=n.s.(.202)$; 1996-$F(3)=0.51$, $p=n.s.(.677)$; 1997-$F(3)=3.49$, $p<.05$; Scheffe 1 x 3 $p<.05$.

Appendix 9: Mean Number of Total Agricultural Tools (excluding tractors) by Number of Non-Redundant Ties

Non-Redundant Ties	1995	1996	1997
0-3	0.00	0.16	0.13
4-5	0.12	0.19	0.23
6	0.14	0.16	0.43
7-8	0.19	0.24	0.29
Total Sample (N=463)	0.14	0.19	0.27

1995-$F(3)=1.07$, $p=n.s.(.364)$; 1996- $F(3)=0.35$, $p=n.s.$; 1997-$F(3)=2.55$, $p=n.s.(.055)$.

Appendix 10: Mean Weighted Number of Animals by Number of Redundant Ties

Redundant Ties	1995	1996	1997
0-8	295.34 (n=132)	324.82 (n=113)	307.60 (n=98)
9-10	359.83 (n=105)	357.67 (n=81)	508.59 (n=66)
11-13	410.34 (n=109)	438.76 (n=108)	450.67 (n=112)
14-16	410.22 (n=117)	387.68 (n=161)	407.25 (n=187)
Total Sample (N=463)	366.07	353.74	411.11

1995-$F(3)=3.63$, $p<.01$, scheffe 1 x 3, $p<.05$, 1 x 4, $p<.05$; 1996-$F(3)=2.06$, p=n.s.(.105); 1997-$F(3)=3.88$, $p<.01$; scheffe 1 x 2 $p<.05$.

Appendix 11: Mean Weighted Number of Animals by Number of Non-Redundant Ties.

Non-Redundant Ties	1995	1996	1997
0-3	251.51 (n=102)	227.74 (n=91)	291.08 (n=89)
4-5	349.50 (n=155)	401.44 (n=145)	360.38 (n=133)
6	438.48 (n=66)	388.18 (n=74)	509.00 (n=)
7-10	441.96 (n=122)	458.72 (n=139)	490.29 (n=144)
11-17	378.00 (n=18)	289.93 (n=14)	316.22 (n=18)
Total Sample (N=463)	366.07	379.00	411.11

1995-$F(3)=6.15$, $p<.001$, Scheffe 0-3 x 6, $p<.01$, 0-3 x 7-10, $p<.001$; 1996-$F(3)=6.61$, $p<.001$, Scheffe, 0-3 x 4-5, $p<.01$, 0-3 x 7-10, $p<.001$; 1997- $F(3)=5.51$, $p<.001$, Scheffe 0-3 x 6, $p<.05$, 0-3 x 7-10, $p<.01$.

Appendix 12: ДОГОВОР АРЕНДЫ ЗЕМЕЛЬНОЙ ДОЛИ

(Agreement for Renting Land Share)
Source: (http://www.igc.apc.org/cci/winpubs/rentland.html)

1. По настоящему Договору собственник земельной доли _____(ФИО) далее именуемый Арендодатель", передает свою долю в аренду, право на которую удостоверено Свидетельством на право собственности на землю №___, выданным "___"_____199__г.

Комитетом по земельным ресурсам и землеустройству _____ района, далее именуемым "Райкомзем", _____(наименование юридического лица, КФХ, ФИО гражданина, ведущего ЛПХ) в лице,_____, (ФИО, должность) действующего на основании _____(вид учредительных документов) далее именуемому "Арендатор", который принимает эту земельную долю на нижеследующих условиях.

2. Настоящий Договор вступает в силу с момента его регистрации в Райкомземе и действует до "___" _____199__ г.(*Срок Договора должен быть не менее чем 3 года.) До истечения указанного срока Стороны могут пересмотреть условия Договора или расторгнуть его по взаимному согласию. Договор после истечения срока его действия автоматически продлевается на год до тех пор, пока от одной Стороны не поступит письменное уведомление другой Стороне о его расторжении не позднее чем за ___ месяцев до истечения срока Договора. Письменное уведомление о расторжении Договора направляется также в Райкомзем.

3. Арендная плата включает:

Виды оплаты, единицы измерения Количество в год (сумма) Сроки (Указать конкретную дату или периодичность платежей.)
1. Деньги (руб.)
2. Продукция и услуги:
3. Налоги и иные платежи за землю (рублей за долю) ежегодно

4. Арендодатель поручает, а Арендатор принимает на себя обязательства по перечислению налоговых и иных платежей за землю в счет арендной платы.

5. Денежные выплаты на каждую дату их осуществления индексируются в соответствии с ростом минимальной заработной платы. При изменении суммы налогов и иных платежей за землю, на ту же сумму изменяется размер арендной платы.

6. Отношения, возникающие в связи с настоящим Договором, регулируются действующим законодательством Российской Федерации.

7. Арендатор имеет право осуществлять от имени Арендодателя действия по выделу в натуре на внутрихозяйственном распределительном собрании земельного участка, соответствующего арендованным земельным долям. Арендодатель признает, что в результате такого выдела образуется участок, принадлежащий на праве общей долевой собственности Арендодателю и другим участникам общей долевой собственности, и что земельная доля в таком участке может измениться по площади и другим характеристикам.

8. Арендатор вправе использовать земельный участок, соответствующий арендованным им земельным долям, только в сельскохозяйственных целях и не имеет права продавать, закладывать или иным образом отчуждать этот земельный участок или земельные доли Арендодателя. После прекращения действия настоящего Договора все права, относящиеся к земельной доле, переходят к Арендодателю.

9. Настоящий Договор составлен в трех экземплярах, имеющих одинаковую юридическую силу, которые выдаются Сторонам по настоящему Договору и Райкомзему. К экземпляру Арендодателя прилагается оригинал свидетельства на право собственности на землю. По просьбе Арендатора к его экземпляру настоящего Договора прилагается копия свидетельства.

10. Настоящий Договор подписан "___" _____ 199__ г. в т _____ (наименование населенного пункта, района, области)

Арендодатель   Арендатор

_____   _____
(ФИО физического лица)   (ФИО физического лица)

паспорт   паспорт

_____   _____
(серия, номер, кем и когда выдан)   (серия, номер, кем и когда выдан)
(наименование юридического лица (КФХ),   (наименование юридического лица (КФХ),
должность представителя, местонахождение, должность представителя, местонахождение,
или ФИО физического лица)   или ФИО физического лица)

адрес   адрес

_____   _____

Договор зарегистрирован в Комитете по земельным ресурсам и землеустройству _____ района "___" _____ 199 ___ г.
Регистрационный номер_____М.П. _____(подпись должностного лица райкомзема)

270 *Household Capital and the Agrarian Problem in Russia*

Appendix 13: Factor Analysis of the Modified CES-D Scale in Three Russian Villages

	Factor loadings in 1995				Factor loadings in 1996				Factor loadings in 1997				Interpretation
	1	2	3	4	1	2	3	4	1	2	3	4	
1. Lonely	**.796**	.179	.115	-.194	**.772**	.458	.164	-.254	**.570**	.333	.168	-.271	Depressed
2. Blues	**.769**	.233	.189	-.236	**.829**	.300	.183	-.306	**.806**	.227	.127	-.376	
3. Effort	**.683**	.376	-.001	-.123	**.535**	.244	-.107	-.009	**.642**	.501	-.008	-.125	
4. Fearful	**.659**	.322	-.003	-.118	**.588**	.008	.005	-.154	**.675**	.149	.165	-.113	
5. Poor appetite	.329	**.755**	.004	-.128	.239	**.732**	-.187	-.201	.214	**.819**	.008	-.192	Somatic-retarded activity
6. Depressed	.262	**.711**	.189	-.239	.007	**.765**	.246	-.216	.234	**.839**	.434	-.324	
7. Could not get going	.507	**.674**	.136	-.001	.385	**.792**	.003	-.178	.292	**.516**	.113	-.222	
8. Dislike	.115	.168	**.960**	-.001	.005	.005	**.951**	-.004	.169	.102	**.939**	-.003	Interpersonal
9. Enjoy	-.008	-.226	-.003	**.825**	-.155	-.158	-.110	**.815**	-.256	-.007	-.136	**.829**	Positive
10. Happy	-.127	-.261	-.001	**.812**	-.009	-.197	.119	**.813**	-.005	-.252	-.009	**.834**	
11. Optimistic	-.296	.102	-.008	**.759**	-.271	-.191	-.104	**.666**	-.315	-.230	-.001	**.677**	

Appendix 14: On-line Information and Data on Agriculture and Rural Russia

1. Ministry of Agriculture and Food of Russia. <http://www.aris.ru/WIN_E/>.
2. The Russian Grain Union. <http://rc.msu.ru/grain/ENGLISH/GRAIN_E.HTM>. The Grain Union is an all-Russia professional organization consisting of grain market participants.
3. Foundation for Support of Agrarian Reform and Rural Development <http://www.raf.org.ru/>.
4. Foundation for Agrarian Development Research. <http://fadr.msu.ru> FADR is a nongovernmental, nonprofit organization whose principal goal is to facilitate the agrarian transformation and sustainable agriculture development in Russia.
5. Practical Science Database. <http://www.sci.aha.ru/cgi-bin/regbase.pl >. Russia in Figures: regional data.
6. The IFC/KHF program. <http://soil.msu.ru/~land> Land Privatization and farm Reorganization in Russia.
7. The Center for Citizen Initiatives. <http://www.igc.org/cci/> CCI is a San Francisco-based non-profit organization dedicated to empowering Russian citizens to take responsibility for their personal futures and assisting Russia in its transition to a market-based economy and a civil society.
8. Rural Enterprise Adaptation Program International. <http://alice.ibpm.serpukhov.su/partners/ccsi/usnisorg/agricltr/reap.htm> REAP's primary objective is to bring efficient, economical and sustainable farming techniques and technology to private farmers in the former Soviet Union.
9. Agricultural, bilingual, English-Russian WWW site dedicated to building bridges between U.S. and NIS farmers and agribusinesses. <http://www.agriculture.com/contents/world/enaindex.html>
10. SovEcon.<http://www.sovecon.ru/wel.html> An independent Moscow-based commodity analyst.
11. Belgorod oblast. <http://www.belgorod.mmtel.ru/> Agriculture.
12. Nizhny Novgorod oblast.<http://www.inforis.ru/n-nov/profile/english/agricult.html>. Agriculture.
13. Samara oblast Commodity Corporation.<http://fpk.online.samara.ru/>.
14. Tula's Market of agricultural products. <http://www.tula.su/businmap/tula/tula.new/torg/>.

Bibliography

Agranaia reforma Rossi. 1998. *Krest'ianskie vedomosti.* March, <http://www.ifc.org.ru/land>.
Agronpromyshlennyi kompleks Belgorodskoi oblasti sostoianie i perspektivy. 1998. <http://www.belgorod.mmtel.ru/>.
APK 1994. "*Programa agronoi reformy v Rossiiskoi Federatsii na 1994-1995 gody,*" *APK: ekonomika, upravlenie,* No. 12 (December): 3-25.
Appels, A., H. B. V. Grabauskas, A., Gostautas and F. Sturmans. 1996. "Self-rated health and mortality in a Lithuanian and Dutch population." *Social Science in Medicine,* 42, 681-689.
Arbuckle, J. L. 1997. *Amos Users' Guide,* Version 3.6. Chicago, IL: SmallWaters Corporation.
Armstrong, P.S. and M. Schulman. 1990. "Financial strain and depression among farm operators: the role of perceived economic hardship and personal control," *Rural Sociology* 55 (3):475-493.
Baily, F.G. 1966. "The peasant view of the bad life." *Advancement in Science* (December): 399-409.
Beggs, J. J., V. A. Haines and J. S. Hurlbert. 1996. "Revisiting the rural-urban contrast: personal networks in nonmetropolitan and metropolitan settings." *Rural Sociology* 61(2): 306-325.
Belyea, M. J. and M. L. Lobao. 1990. "Psychosocial consequences of agricultural transformation: the farm crisis and depression," *Rural Sociology* 55 (1):55-75.
Benet, S. 1970. *The Village of Viriatino: An Ethnographic Study of a Russian Village from Before the Revolution to the Present.* New York: Doubleday Anchor Books.
Boyd, J. H., M. M. Weissman, W. D. Thompson and J.K. Myers. 1982. "Screening for depression in a community sample: understanding the discrepancies between depression symptoms and diagnostic scales." *Archives of General Psychiatry* 39:1195-1204.
Campbell, A. 1981. *The Sense of Well-Being in America.* New York: McGraw-Hill.
Cambell, A, P. Converse and W. Rodgers. 1976. *The Quality of American Life: perceptions, evaluations and satisfaction.* New York: Russell Sage Foundation.
Center for Russian Studies. 1998. "Database." NUPI. <http://www.nupi.no/russland/russland.htm>.
Chaianov, A.V. 1966. *The Theory of Peasant Economy.* Homewood, IL: R.D. Irwin.
Ciarlo, J. A. and D. L. Tweed. 1992. "Exploring Colorado's need for mental health servcies: some preliminary findings." Paper presented to the meeting of the American Evaluation Association, Seattle, WA.

Coleman, J. S. 1988. "Social capital in the creation of human capital." *American Journal of Sociology* 94 (special supplement): S95-120.
Danes, S. M., O. N. Doudchenko and L. V. Yasnaya. 1994. "Work and family life." Pg. 156-185 in James W. Maddock, M. Janice Hogan, Anatoly I. Oatonov and Mikhail S. Matskovsky (eds.), *Families Before and After Perestroika: Russian and U.S. Perspectives.* New York: Guilford Press.
Decree of President, RF. 1993. "*O regulirovanii zemykh otnoshenii i razvitii agranoii reformi.*" No. 1761, October 27th.
Decree of President, RF. 1996. "*O realizatsii konstitutsonnkh prav grazhdan na zemlu.*" No. 337, March 7th.
Deere, C. D. and A. de Janvry. 1981. "Demographic and social differentiation among Northern Peruvian peasants." *The Journal of Peasant Studies* 8:335-366.
DeForge, B. R. and J. Sobal. 1988. "Self-report depression scales in the elderly: the relationship between the CES-D and Zung." *International Journal of Psychiatry in Medicine* 18(4):325-338.
Dershem, L. D. 1998. "Prevelance, sources and types of informal support in Latonovo and Mayaki." Pp. 163-198 in David J. O'Brien, Valeri V. Patsiorkovski, Larry D. Dershem, Alessandro Bonanno and Charles Timberlake, *Services and Quality of Life in Rural Villages in the Former Soviet Union: Data from 1991 and 1993 Surveys.* Lanham, Maryland: University Press of America.
Dershem, L. D. 1995. *Community and Collective: interpersonal ties in three Russian villages.* Unpublished Ph.D. dissertation, Columbia, MO: University of Missouri.
Dershem, L. and D. Gzrishvili. 1998. "Informal social support networks and household vulnerability: empirical findings from Georgia." *World Development* Vol.26 (10):1827-1838.
Dershem, L. and V. Patsiorkovski. 1997. *The Needs and Capacities of the Third Sector in Eight Oblasts of Central Russia.* Report to Save the Children Federation, Moscow, Russia.
Dershem, L., V. Patsiorkovski and D. O'Brien. 1996. "The use of the CES-D for measuring symptoms of depression in three rural Russian villages." *Social Indicators Research* 39:89-108.
Ekonomicheskaia Gazeta. 1994. *Ekonomicheskaia Gazeta.* Number 39: 18.
Ensel, W. M. 1986. "Measuring depression." Pp. 51-70 in *Social Support, Life Events, and Depression*, edited by N. Lin, A. Dean and W. Ensel. Orlando, FL: Academic Press.
Fischer, C. S. 1982. *To Dwell Among Friends: personal networks in town and city.* Chicago: University of Chicago.
Fitzpatrick, S. 1994. *Stalin's Peasants: resistance and survival in Russian villages after collectivization.* New York: Oxford University Press.
Flora, C. B. and J. L. Flora. 1993. "Entrepreneurial social infrastructure: A necessary ingredient." *Annals, AAPS* 529:48-58.

Foster, G. M. 1965. "Peasant society and the Image of the limited good." *American Anthropologist* Vol. 67(2):293-315.
Foundation for Support of Agrarian Reform and Rural Development. 1998. "*Programma privatizatsii zemli i reorganizatsii sel'skokhoziaistvennykh predpriiatii.*" <http://www.raf.org.ru.>.
Fugita, S. S. and D. J. O'Brien. 1991. *Japanese American Ethnicity: the persistence of community.* Seattle: Washington University Press.
Gatz, M. and M. Hurwicz. 1990. "Are old people more depressed? Cross-sectional data on Center for Epidemiological Studies of depression scale factors." *Psychology and Aging*, 5: 284-290.
Goskomstat. 1990. *Narodnoe khoziaistvo Rossiiskoi Federatsii v 1989.* (Households in the Russian Federation in 1989) Moscow: Goskomstat.
Goskomstat. 1991. *Sel'skie naselennye punkty* RSFSR. Moscow: Goskomstat.
Goskomstat. 1992. *Chislennost' sostav i dvizhenie naseleniia v Rossiiskoi Federatsii.* Moscow: Goskomstat.
Goskomstat. 1994. *Tipy issotav domokhosaistv v Rossii po mikroperepisi 1994 goda.* Moscow: Goskomstat.
Goskomstat. 1995. *Rossiiski statisticheskii ezhegodnik.* (Russian statistical yearbook) Moscow: Goskomstat.
Goskomstat. 1996. *Rossiiski statisticheski ezhogodnik.* (Russian statistical yearbook) Moscow: Goskomstat.
Goskomstat. 1997a. *Demograficheskii ezhegodnik Rossii* (The Demographic Yearbook of Russia 1997). Moscow: Goskomstat.
Goskomstat. 1997b. *Rossiia v tsifrkh* 1997. Moscow: Goskomstat.
Goskomstat. 1997. *Chislennost' naselenia Rossiiskoi Federatsii po polu i vozrastu na* 1 January. 1997. Moscow: Goskomstat.
Goudy, W. 1990. "Community attachment in a rural region." *Rural Sociology* 55:178-98.
Granovetter, M. S. 1973. "The strength of weak ties." *American Journal of Sociology* 78:1360-1380.
Himmelfarb, S. and S. A. Murrel. 1983. "Reliablity and validity of five mental health scales in older persons." *American Journal of Gerontology* 117:173-183.
Hough, J. F. 1994. "The Russian election of 1993: Public attitudes toward economic reform and democratization." *Post-Soviet Affairs* 10(1):1-37.
Hough, J. F., E. Davidheiser and S. Goodrich-Lehmann. 1996. *The 1996 Russian Presidential Election.* Washington, D.C.: The Brookings Institution.
Hoyt, D. R. 1988. *Economic stress and mental health in rural Iowa.* Unpublished manuscript.
Hoyt, D. R., D. O'Donnell and K.Y. Mack. 1995. "Psychological distress and size of place: the epidemiology of rural economic stress." *Rural Sociology* 60 (4): 707-720.
Humphrey, C. 1988. "Rural society in the Soviet Union." pp. 53-70 in *Understanding Soviet Society,* edited by Michael P. Sacks and Jerry G. Pankhurst. Boston: Allen and Unwin.

Husaini, B.A., J.A. Neff, J.B. Harrington, M.D. Hughes and R.H. Stone. 1980. "Depression in rural communities: validating the CES-D scale," *Journal of Community Psychology* 8: 20-27.
IFC (International Financial Corporation). 1996. Agrarian reform in Russia. <http://soil.msu.ru/~land>
Kadushin, C. 1983. "Mental health and the interpersonal environment." *American Sociological Review* 48:188-98.
Karcz, J. F. 1971. "From Stalin to Brezhnev: Soviet agricultural policy in historical perspective." Pp. 36-72 in James R. Millar (ed.), *The Soviet Rural Community.* Urbana: University of Illinois Press.
Kasarda, J. D. and M. Janowitz. 1974. "Community attachment and mass society." *American Sociological Review* 48:328-339.
Kessler, R. C. and D. F. Greenberg. 1981. *Linear Panel Analysis*: Models of Quantitative Change. New York: Academic Press.
Klugman, J. 1997. *Poverty in Russia: public policy and private responses.* The World Bank: Washington, D.C.
Kohout, F. J., L. F. Berkman, D. A. Evans and J. Cornoni-Huntley. 1993. "Two shorter forms of the CES-D depression symptoms index." *Journal of Aging and Health* 5(2):179-193.
Kornhauser, W. 1959. *The Politics of Mass Society.* New York: The Free Press.
Koznova, I. E. 1996. "*Sotsial'naia pamiat' krest'ianstva kak faktor.*" <http://www.fadr.msu.ru/archives/mailing-list/priv-agr/art-rus/msg00021.html>.
Light, I. H. 1972. *Ethnic Enterprise in America.* Berkeley: University of California Press.
Lin, N., A. Dean and W. Ensel. 1986. *Social Support, Life Events and Depression.* Orlando: Academic Press.
Linn, J. G. and B. A. Husaini. 1987. "Determinants of psychological depression and coping behaviors of Tennessee farm residents." *Journal of Community Psychology* 15: 520-536.
Maddox, G.L. and E.B. Douglass. 1973. "Self-assessment of health: a longitudinal study of elderly subjects." *Journal of Health and Social Behavior* 14: 87-93.
Martynov A. S., Artukhov V. V., Vinogradov V. G. 1997. *Rossia kak sistema. Rossia v Tsifrakh. Regionalny spravochnik.* Moscow: Practical Science, 1997. <http://www.sci.aha.ru/RUS/wash_html>.
Martynova, I, K. 1996. *voprosu o realnykh sotsialnykh izmeneniiakh I otsenke sovremennykh preobrazovanii v agrarnom sektore Rossii.* "International Electronic Conference on Privatisation and agrarian Reform in Russia." March-July, 1996. Moscow: WWW server – FIAR, 1996. <http://www.fadr.msu.ru/archives/mailing-list/priv-agr/art-rus/msg00024.html>.
McClendon, M. J. and D. J. O'Brien. 1988. "Question-order effects on the determinants of subjective well-being." *Public Opinion Quarterly* 52:351-364.

Mirowsky, J. and C. E. Ross. 1989. *Social Causes of Pyschological Distress.* New York: Aldine.
Mossey, J. M., and E. Shapiro. 1982. "Self-rated health: a predictor of mortality among the elderly." *American Journal of Public Health*, 72: 800-808.
Narodnoe Khoziaistvo Rossiiskoi 1993. 1993. *Narodnoe Khoziaistvo Rossiiskoi Federatsii v 1992.* Moscow: Informatsionno-izdatel'skii Tsentr.
Nazarenko, V. I. 1998. "*Sel'skoe khoziastvo Rossii I mirovoi rynok.*" <http://www.fadr.msu.ru/archives/mailing-list/priv-agr/art-rus/msg00004.html>.
Neff, J. A. and B. A. Husaini. 1987. "Urbanicity, race, and psychological distress." *Journal of Community Psychology* 15:520-536.
Netting, R. 1993. *Smallholders, Householders: Farm Families and the Ecology of Intensive, Sustainable Agriculture.* Stanford, California: Stanford University Press.
Nezavisimaya Gazeta. 1996. *Nezavisimaya Gazeta.* No. 109 (1188). 18 (June): 3.
Nicolsky C. A. 1996. "*Agrarnaia reforma* 1991-1996 *godov i problema modernizatsii russkoi derevni..*" *International Electronic Conference on Privatisation and agratian Reform in Russia.* March-July, 1996. Moscow: WWW server – FIAR.
North, D. C. 1990. *Institutions, Institutional Change and Economic Performance.* Cambridge: Cambridge University Press.
O'Brien, D. and S. S. Fugita. 1991. *The Japanese American Experience.* Bloomington, Indiana: Indiana University Press.
O'Brien, D. J., E. W. Hassinger, R. B. Brown and J. R. Pinkerton. 1991. "The social networks of leaders in more and less viable rural communities." *Rural Sociology* 56:699-716.
O'Brien, D. J., V V. Patsiorkovski, I. Korkhova and L. D. Dershem. 1993. "The Future of the Village in a Restructured Food and Agricultural Sector in the Former Soviet Union." *Agriculture and Human Values* 10:13-22.
O'Brien, D. J., E. W. Hassinger and L. Dershem. 1994. "Community attachment and depression among residents in two rural Mid-Western communities." *Rural Sociology* 59 (2): 255-265.
O'Brien, D. J., V. V. Patsiorkovski, L. D. Dershem and O. Lylova. 1996a. "Social Capital and Adaptation to Social Change in Russian Villages." Studies in Public Policy. Centre for the Study of Public Policy. University of Strathclyde. Glasgow, Scotland.
O'Brien, D. J., V. V. Patsiorkovski, L. D. Dershem and O. Lylova. 1996b. "Peasant household production and symptoms of stress in post-Soviet Russian villages." *Rural Sociology* 61(4):674-698.
O'Brien, D. J., E. W. Hassinger and L. Dershem. 1996c. "Size of place, residential stability and personal social networks." *Sociological Focus* 29(1):61-72.

O'Brien, D. J., V. V. Patsiorkovski and L. D. Dershem. 1998a. "Rural responses to land reform in Russia: An analysis of household land us in Belgorod, Rostov and Tver' *Oblasts* from 1991 to 1996." pp. 35-61 in Stephen K. Wegren (editor), *Land Reform in the Former Soviet Union and Eastern Europe.* London: Routledge.

O'Brien, D. J., V. V. Patsiorkovski, L. D. Dershem, A. Bonanno and C. Timberlake. 1998b. *Services and Quality of Life in Rural Villages in the Former Soviet Union.* Lanham, Maryland: University Press of America.

O'Brien, D. J., A. Raedeke and E. W. Hassinger. 1998c. "The social networks of leaders in more and less viable communities six years later: A research note." *Rural Sociology* 62 (3):109-127.

O'Hara, M., F. Kohout and R. Wallace. 1985. "Depression among the rural elderly: a study of prevalence and correlates." *The Journal of Nervous and Mental Disease* 173(10): 582-589.

Olson, M. Jr. 1971. *The Logic of Collective Action Public Goods and The Theory of Groups.* Cambridge, Mass: Harvard University Press.

Patsiorkovski, V. V. 1991. *The Paid Services of the Population.* Moscow: Nauka.

Patsiorkovski, V. V., A. Bonanno, J. Chinn, and D. J. O'Brien. 1991. "Selected rural issues in the USA and the USSR: A comparative agenda." *The Rural Sociologist* 11:21-35.

Patsiorkovski, V.V and N. Rimashevskaya. 1991. *People's Well-Being: trends and prospects.* Moscow: Nauka.

Patsiorkovski, V.V. and D. J. O'Brien. 1996. *Research Methodology and Quality of Life in Russia and the USA.* Moscow-Columbia: ISESP, RAS.

Pearlin, L. I. 1989. "The sociological study of stress." *Journal of Health and Social Behavior* Vol. 30 (September):241-256.

Popov, N. 1996. "*Krest'iamskie (fermerskie) khoziaistva.*" *APK, Ekonomika, Upravlenie,* No. 5 (May).

Proggramma privatizatsii i reorganiizatsii sel'skokhoziaistvennykh predpriiatii. 1998. Moscow: *Foundation for Support of Agrarian Reform and Rural Development.* <http://www.raf.org.ru>.

Prosterman, R. L. and T. Hanstad. 1993. "A fieldwork-based appraisal of individual peasant farming in Russia." pp. 149-192 in Don Van Atta (ed.) *The "Farmer Threat": the political economy of agrarian reform in post-Soviet Russia.* Boulder, CO.: Westview Press.

Putnam, R. D. 1993. "The prosperous community: Social capital and public life." *The American Prospect* 13:35-42.

Radloff, L. 1977. "The CES-D scale: a self-report depression scale for research in the general population." *Applied Psychological Measurement* 1(3):385-401.

Rimashevskaya, N. 1997. "Poverty trends in Russia: a Russian perspective". pp. 119-131 in *Poverty in Russia: public policy and private responses,* edited by Jeni Klugman. Washington, D.C.: The World Bank.

Roberts, R. E. 1980. "Reliability of the CES-D scale in different ethnic contexts." *Psychiatry Research,* 2: 125-134.

Rose, R. and I. McAllister. 1996. "Is money the measure of welfare in Russia?" *Review of Income and Wealth*, Series 42, No. 1 (March): 1-16.
Ross, C. E. and J. Huber. 1985. "Hardship and depression." *Journal of Health and Social Behavior* Vol. 26 (December): 312-327.
Rossiia, 1997. 1998. *Seventh Annual Report*. Moscow: Institute for the Socio-Economic Studies of Population, Russian Academy of Sciences.
Rossiiskaya Gazeta. 1996. September 21.
Russia in Figures, 1997. 1997. Moscow: Goskomstat.
Salamon, S. 1985. "Ethnic communities and the structure of agriculture." *Rural Sociology* 50:323-340.
Sanders, J. M. and V. Nee. 1998. "Immigrant self-employment: the family as social capital and the value of human capital." *American Sociological Review* Vol. 16 (April): 231-249.
Schulman, M. D. and P. S. Armstrong. 1989. "The farm crisis: an analysis of social psychological distress among North Carolina farm operators." *American Journal of Community Psychology* Vol. 17, No. 4: 423-439.
Sel'skie naselennye punkty RSFSR. Moscow: Goskomstat RSFSR, 1991.
Selskaya zhizn. 1996a. *Selskaya zhizn*. June 6.
Selskaya zhizn. 1996b. *Selskaya zhizn*. May 23.
Selskaya zhizn. 1996c. *Selskaya zhizn*. March 12.
Selskaya zhizn. 1996d. *Selskaya zhizn*. March 14.
Selskaya zhizn. 1996e. *Selskaya zhizn*. February 24.
Semenov, V.A. 1998. "*Ob itogakh raboty v pervom polugodee* 1998," Ministry of Agriculture, Russian Federation: Ufa June 23. <http://www.aris.ru/N/WINR/PRESS/MINISTR/dokl-ufa.html>.
Schumpeter, J. A. 1950. *Capitalism, Socialism and Democracy*. Third Edition. New York: Harper and Row.
Shanin, T. 1982. "Defining peasants: conceptualisations and de-conceptualisations." *The Sociological Review* 30:407-431.
Shearer, E. and F. Starr. 1996. "Through a prism darkly." *American Journalism Review*: 37-40.
Shlapentokh, V. 1989. *Public and private life of the Soviet people : changing values in post-Stalin Russia*. New York: Oxford University Press, 1989.
Smirnov V.V. 1996. "International Electronic Conference on Privatization and agrarian Reform in Russia." *Professionalnaya sotsialnaya rabota*. March-July, Moscow:<http://www.fadr.msu.ru/archives/mailing-list/priv agr/artrus/msg00026.html>.
Smith, H. 1990. *The New Russians*. New York: Random House.
Tausig, M. 1986. "Measuring life events." pp. 71-95 in *Social Support, Life Events, and Depression* edited by Nan Lin, Alfred Dean and Walter Ensel. New York: Academic Press, Inc.
The Economist. 1998. July 25th.
Tocqueville, A. de. 1945. *Democracy in America*. Two Volumes. New York: Alfred A. Knopf, Vintage Books. Originally published in 1835.

Tuma, N. B. and M. T. Hannan. 1984. *Social Dynamics: Models and Methods*. Orlando: Academic Press.

Turner, R., B Wheaton and D. Lloyd. 1995. The epidemiology of social stress. *American Sociological Review*, Vol. 60 (February): 104-125.

Tipy i sostav domokhosaistv v Rossii po mikroperepisi 1994 goda. 1995. Moscow: Goskomstat.

Uroven' Gizni Naselenia Rossiiskoi Federatsii. 1995. Moscow. Goskomstat Rossii.

U. S. Department of Commerce 1984. *1982 Census of Agriculture*. U. S. Department of Commerce, Bureau of the Census, Washington, D. C., Vol. 1, Part 25.

U. S. Department of Commerce 1994. *1992 Census of Agriculture*. U. S. Department of Commerce, Bureau of the Census, Washington, D. C., Vol. 1, Part 25.

Van Atta, D. (ed.). 1993. *The "Farmer Threat": the political economy of agrarian reform in post-Soviet Russia*. Boulder: Westview Press.

Vinogradsli, V. 1997. *Baiat'-znachit' govorit' (golosa krestian Rossii)*. Electronic conference *Fonda Issledovania Agrarnogo Razvitia* (FIAR) FADRNEWS. Archive FADRNEWS za 1996-1998, Moscow: <http://www.fadr.msu.ru/mailserv/fadrnews/msg00083.html>.

Vishnevsky, A. G. 1996. "Family, fertility, and demographic dynamics in Russia: analysis and forecast." *Russia's Demographic "Crisis"*. Edited by Julie DaVanzo. Rand Corporation. <http://www.rand.org/publications /CF/CF124/CF124.chap1.html>.

Wegren, S. K. 1998. "The conduct and impact of land reform in Russia." pp. 3-34 in Stephen K. Wegren (editor), *Comparative Land Reform in the Former Soviet Union*. London: Routledge.

Wegren, S. K. and V. R. Belen'kiy. 1998. "The political economy of the Russian land market." *Problems of Post-Communism* (July/August): 56-66.

Weissman, M. and G. Lerman. 1977. "Sex difference in the epidemiology of depression." *Archives of General Psychiatry*. 34: 98-111.

Wellman, B. and S. Wortley. 1990. "Different strokes from different folks: community ties and social support." *American Journal of Sociology*, No. 3 (November): 558-588.

White, S., R. Rose and I. McAllister. 1997. *How Russia Votes*. Chatham, New Jersey: Chatham House Publishers.

Wilson, K. L. and A. Portes. 1980. "Immigrant enclaves: an analysis of the labor market experiences of Cubans in Miami." *American Journal of Sociology* 86: 295-319.

Yaney, G. L. 1971. "Agricultural administration in Russia from the Stolypin land reform to forced collectivization: An interpretive study." pp. 3-35 in James R. Millar (ed.) *The Soviet Rural Community*. Urbana: University of Illinois.

Ying, Y. 1988. "Depressive symptomology among Chinese-Americans as measured by the CES-D." *Journal of Clinical Psychology* 44 (5):79-746.

Index of Authors

Appels et al., 201
Arbuckle, 76, 143
Armstrong and Schulman, 193
Baily, 193
Beggs et al., 93
Belyea and Lobao, 194, 196
Benet, 69
Boyd et al., 194
Campbell, 74
Campbell, Converse and
  Rodgers, 214, 215
Chaianov, 26, 72, 144
Ciarlo and Tweed, 194
Coleman, 26, 31, 159
Danes et al., 69
Deere and de Janvry, 26, 72, 144
Deforge and Sobal, 194
Dershem, 32, 40, 62, 72, 73, 84
Dershem and Gzirishvili, 37
Dershem and Patsiorkovski, 245
Dershem et al., 62, 74, 94, 192, 194, 197
Ensel, 74, 196, 200
Erlanger, 10
Fischer, 72, 73, 93
Fitzpatrick, 27, 28
Flora and Flora, 31
Foster, 193
Fugita and O'Brien, 25, 31
Gatz and Hurwicz, 192, 201
Goudy, 95
Granovetter, 94, 160
Hanstad, 16

Himmelfarb and Murrell, 194
Hough, 17
Hoyt, 194
Humphrey, 20, 28, 32
Husaini and Neff, 194
Husaini et al., 196
Kadushin, 93
Karcz, 16
Kessler and Greenberg, 76
Klugman, 181
Kohout et al., 194
Kornhauser, 31
Koznova, 97
Light, 25, 31
Lin et al., 39, 191
Linn and Husaini, 194
Maddox and Douglas, 201
Martynov et al., 52
Martynova, 43
McClendon and O'Brien, 74
Mirowsky and Ross, 73
Mossey and Shapiro, 201
Nazarenko, 121
Neff and Husaini, 194
Netting, 25
Nicolsky, 43
North, 1, 115, 239
O'Brien, 2, 16
O'Brien and Fugita, 115
O'Brien et al., 2, 16, 18, 20, 22, 23, 24, 27, 29, 31, 32, 34, 39, 62, 64, 66, 67, 69, 71, 72, 73, 74, 85, 93, 94, 101, 111, 115, 151, 192, 194, 196, 239, 243

O'Hara et al., 195, 206
Olson, 137
Patsiorkovski, 27, 68
Patsiorkovski and O'Brien 2, 20, 22, 29, 30, 32, 62, 65, 69, 85
Patsiorkovski and Rimashevskaya, 68
Patsiorkovski et al., 24, 94
Pearlin, 201
Popov, 18
Prosterman, 16
Putnam, 1
Radloff, 73, 193
Rimashevskaya, 182
Roberts, 193, 194
Rose and McAllister, 37, 164
Ross and Huber, 38, 193
Salamon, 238
Sanders and Nee, 25
Schulman, 193
Schumpeter, 240

Semenov, 6
Shanin, 20
Shearer, 10
Slapentokh, 32
Smirnov, 57
Smith, 16
Starr, 10
Tausig, 201
Tocqueville, 1
Tuma and Hannan, 76
Turner et al., 193
Van Atta, 16, 28, 34
Vinogradsli, 43
Vishnevsky, 44
Wegren, 8, 16, 17, 18, 24, 30, 43, 57, 82, 115, 141
Wegren and Belen'kiy, 18
Weissman and Lerman, 206
White et al., 213
Wilson and Portes, 25, 31, 160
Yaney, 16
Ying, 194

Subject Index

A

Acquisition of Durable Goods, 36, 37
agricultural equipment, 34, 109, 119, 129, 159
agricultural sales, 11, 13, 71, 77, 151, 153, 155, 167, 181, 228
American Midwest, 30, 39, 73, 85, 91, 107, 115, 192
Amos, 76, 273
automobiles, 10, 62, 120, 129, 183, 185, 187, 189

B

birth rates, 57
Book of Household Accounts, 68

C

Center for Epidemiological Study of Depression Scale. *See* CES-D
CESD, 62
chemical fertilizers, 21, 120
Chernomyrdin, 17
chernozem, 19, 63, 64, 79, 82, 92, 97, 107, 111, 133
Churches, 96
civic culture, 106, 160, 178, 181
civic society, 1, 190, 218
civic voluntary organizations, 245

collective farms, 64, 65, 66
collective goods, 26, 80
commercial credit markets, 243
communal land, 72, 114, 115
Communist party, 17
community attachment, 22, 23, 31, 33, 34, 35, 36, 37, 39, 40, 41, 62, 77, 92, 93, 95, 96, 104, 131, 192, 199, 207, 224, 226
community development, 31, 245
community involvement, 34, 73, 93, 99, 104, 106, 107, 117, 123, 129, 130, 132, 140, 148, 153, 155, 158, 159, 171, 172, 175, 177, 178, 180, 181, 187, 189, 190, 213, 220, 224, 226, 228, 230, 234, 236, 239
conservatism, 237
contracts, 26, 115
cooperatives, 239, 242, 244
Czech Republic, 242

D

democratic institutions, 16, 108
divorce, 83, 84
divorced, 70, 83, 86, 201, 205, 221, 230
Duma, 6, 16, 109

E

economic viability, 170
elderly, 56, 57, 58, 59, 80, 81, 82, 83, 85, 88, 112, 118, 119, 137, 147, 153, 161, 170, 171, 172, 177, 178, 187, 192, 197, 200, 228, 230, 234, 242, 274, 276, 277, 278
emerging niche markets, 23
employed couples with children, 69, 81, 82, 83, 84, 89, 90, 117, 209
employed couples with children and other adults, 69, 90, 117
employed couples without children, 69, 81
employment, 3, 12, 65, 69, 74, 81, 138, 161, 162, 163, 164, 189, 215, 279
extended family households, 29

F

family festivals, 97, 98
farm reorganization, 19
fatalism, 193
fermery, 8, 18, 19, 20, 24, 75, 111, 114, 163, 238
financial capital, 66, 124
free-market agriculture, 1

G

German Catholic farmers, 238

H

herbicides and pesticides, 121
homeless, 11

household capital, xxi, 13, 22, 33, 40, 41, 43, 77, 131, 143, 144, 155, 161, 171, 177, 178, 181, 213, 214, 215, 220, 224, 230, 235, 239, 240
household economic enterprises, 12
household enterprises, 12, 13, 40, 164, 167, 168, 170, 171, 172, 175, 177, 178, 181, 183, 184, 189, 213, 217, 220, 230, 234, 238, 242, 243, 244
household income, 11, 28, 38, 161, 167, 168, 171, 172, 175, 177, 178, 180, 182, 183, 184, 185, 200, 202, 204, 205, 207, 220, 230, 234
household labor, 11, 22, 23, 24, 25, 27, 33, 35, 36, 37, 38, 39, 40, 41, 43, 47, 51, 52, 54, 57, 58, 59, 60, 68, 72, 76, 77, 79, 88, 89, 90, 91, 92, 104, 106, 108, 116, 122, 127, 129, 130, 131, 132, 136, 137, 138, 139, 141, 144, 147, 152, 153, 159, 171, 172, 178, 184, 187, 189, 190, 191, 192, 198, 202, 208, 209, 213, 220, 221, 224, 228, 230, 235, 239
household plots, 3, 12, 21, 34, 39, 112, 114, 167
household production, 69
household survival strategies, 230

I

inequalities, 41, 240
informal marketing, 32, 175
informal social support, 32
inmigration, 57, 58

Subject Index 285

institutional changes, 2, 41, 235, 242

K

kinship networks, 84
kniga ucheta domashnih khoziaistva. See Book of Household Accounts
kolkhoz, 65, 66
kolkhozy, 3, 6, 9, 12, 16, 17, 19, 20, 27, 28, 29, 34, 39, 40, 62, 64, 75, 104, 109, 110, 161, 189, 238
krestianskoe khoziastva. See peasant farmer
krestianskoe khoziastvo. See peasant household

L

land market, 2, 16, 17, 35, 141, 244, 280
land reform, 16, 278, 280
leadership, 40, 41, 61, 87, 127
legal support, 106
life domains, 74, 77, 215, 216, 217, 220, 221, 224, 226, 234, 235
life expectancy, 58, 59, 69, 84
limited good, 193, 275
livestock, 28, 109, 123, 126, 127, 129, 130, 131, 133, 148, 154, 158, 159
local government, 3, 8, 18, 40, 110, 112, 114, 151, 163, 165, 189, 244

M

marital status satisfactions, 216

material infrastructure, 3, 12, 29, 243, 244, 245
mechanical equipment, 34
medical services, 27
mental health, xxi, 39, 41, 43, 62, 73, 74, 93, 99, 101, 160, 192, 199, 224, 273, 275
migration, 30, 45, 48, 57, 58, 59, 68, 82
monetized income, 36, 71, 131, 164, 168, 182, 183
motorcycles, 119, 120

N

negative life events, 201
neighbors, 11, 27, 28, 35, 36, 112, 115, 117, 127, 130, 140, 144, 158, 167, 172, 175, 185, 239
network analysis, xxi, 94
New Economic Policy, 27
non-agricultural businesses, 167
non-governmental associations, 4
nonmonetized, 36, 37, 71, 131, 164, 166, 167, 182, 183
not-for-profit organizations, 218

O

obiazatel'naia postavka. (production quotas)
other extended family, 69

P

physical capital, 11, 13, 23, 71, 109, 110, 128, 130, 132, 141, 142, 144, 155, 158, 159, 178, 238, 239
podzol, 63, 64, 67, 79, 83, 107
positive life events, 202, 208
post office, 65, 66, 67
poverty, 36, 158, 181, 182, 183

poverty line, 182, 183
private plots, 65, 69
privatization, 9, 15, 17, 32, 43, 62, 112, 114, 115
protectionism, 237
provincial government, 171
psychological distress, 39, 191, 214, 277, 279

Q

quasi-kin, 31

R

rental of land, 112, 117, 123, 129, 131, 136, 141, 144, 148, 151, 159
retired couples, 29, 69, 80, 81, 117, 137, 140, 171, 172, 208, 209
rynok, 22, 277

S

satisfaction with income, 217, 220, 224, 230
school, 83, 91, 96, 126, 161
Sel'soviet. See . See . See
semantic differential items, 74, 216
sense of fitting into the community, 104, 234
single adult household, 69
single parents with children, 69
smallholder farming, 144
social capital, xxi, 1, 2, 11, 12, 13, 15, 16, 22, 25, 26, 31, 32, 34, 35, 38, 39, 41, 43, 61, 62, 63, 64, 66, 68, 72, 79, 80, 81, 85, 91, 92, 93, 99, 102, 104, 106, 108, 110, 127, 128, 129, 130, 132, 136, 137, 141, 142, 147, 153, 155, 156, 158, 159, 160, 161, 171, 175, 178, 180, 183, 187, 189, 190, 191, 192, 202, 206, 207, 209, 211, 213, 224, 234, 235, 236, 237, 238, 279
social exchange helping networks, 23, 31, 41, 62, 92, 93, 101, 107, 129, 159, 213, 214, 234
social service safety net, 213
social services, 69
social welfare, 242, 245
sovkhoz. See state farm. See state farm. See state farm
sovkhozy, 3, 6, 9, 12, 16, 17, 19, 20, 27, 28, 34, 62, 64, 109, 110
subjective quality of life, xxi, 13, 38, 39, 41, 43, 61, 62, 74, 94, 101, 102, 194, 214, 215, 221, 235
subsidies, 6, 16, 28, 237
sustainable agriculture, 2, 43, 271
symptoms of depression, 13, 38, 77, 192, 193, 196, 197, 198, 200, 202, 205, 206, 207, 208, 210, 211, 221, 224, 228, 274

T

Taganrog, 65
telephones, 10, 71, 183, 187
transaction costs, 26, 39, 132, 191, 192
truck farming, 23
trust, 2, 25, 26, 61, 114, 137, 244

U

United States, 23, 31, 38, 67, 85, 95, 116, 164, 244
urban dwellers, 8, 18, 238

V

value added products, 23
VCRs, 10, 62, 71, 183, 184, 187, 189
village ceremonies, 96
Village Council, 34, 67, 68, 72
village satisfaction, 216, 220, 224, 226, 231, 234
voluntary associations, 32
vulnerable persons, 140

W

well-being, xx, 36, 201, 238, 276
Western Europe, 67, 164
women, 54, 69, 70, 84, 85, 86, 161, 162, 164, 192, 206, 210, 212
working-age adults, 35, 36, 38, 66, 142, 162, 164
world economic system, 237

Y

Yeltsin, 6, 17, 112

Z

Ziuganov, 17